SpringerBriefs in Electrical and Computer Engineering

Control, Automation and Robotics

Series editors

Tamer Başar
Antonio Bicchi
Miroslav Krstic

More information about this series at http://www.springer.com/series/10198

Alessandro N. Vargas · Eduardo F. Costa
João B.R. do Val

Advances in the Control of Markov Jump Linear Systems with No Mode Observation

 Springer

Alessandro N. Vargas
Electrotechnical Department
Universidade Tecnológica Federal do
  Paraná, UTFPR
Cornelio Procópio, Paraná
Brazil

João B.R. do Val
FEEC
Universidade Estadual de Campinas
Campinas, São Paulo
Brazil

Eduardo F. Costa
Departmento de Matemática Aplicada e
  Estatística
Universidade de São Paulo, USP
São Carlos, São Paulo
Brazil

ISSN 2191-8112   ISSN 2191-8120 (electronic)
SpringerBriefs in Electrical and Computer Engineering
ISSN 2192-6786   ISSN 2192-6794 (electronic)
SpringerBriefs in Control, Automation and Robotics
ISBN 978-3-319-39834-1   ISBN 978-3-319-39835-8 (eBook)
DOI 10.1007/978-3-319-39835-8

Library of Congress Control Number: 2016940333

© The Author(s) 2016
This work is subject to copyright. All rights are reserved by the Publisher, whether the whole or part
of the material is concerned, specifically the rights of translation, reprinting, reuse of illustrations,
recitation, broadcasting, reproduction on microfilms or in any other physical way, and transmission
or information storage and retrieval, electronic adaptation, computer software, or by similar or dissimilar
methodology now known or hereafter developed.
The use of general descriptive names, registered names, trademarks, service marks, etc. in this
publication does not imply, even in the absence of a specific statement, that such names are exempt from
the relevant protective laws and regulations and therefore free for general use.
The publisher, the authors and the editors are safe to assume that the advice and information in this
book are believed to be true and accurate at the date of publication. Neither the publisher nor the
authors or the editors give a warranty, express or implied, with respect to the material contained herein or
for any errors or omissions that may have been made.

Printed on acid-free paper

This Springer imprint is published by Springer Nature
The registered company is Springer International Publishing AG Switzerland

Contents

Preliminaries

1 Introduction

In the past two decades, there has been a growing interest in the class of stochastic systems known as *Markov jump linear systems* (MJLS). MJLS has received attention due to its capability of modeling processes subject to abrupt variations—examples spam in the literature, such as in the control of paper mills [1], robotics [2, 3], economy [4, 5], networks [6], just to cite a few.

A strong research community has emerged over the last years, with scholars improving progressively the knowledge about MJLS; for instance, the books [7], [8], and [9] are key references on the topic. Besides, papers containing results for stability, control, and other properties of MJLS can be found in the literature; for a brief account, see [1, 2, 4, 5, 10–27].

Notwithstanding the large number of contributions on control of MJLS, most of the available results deal with the case in which the controller has complete access to the Markov state. Even in the context of partial information, most results focus on the jump mode observation, see for instance [28, 29]. In practice, this signifies that the controller has a built-in sensor or a similar measurement instrument that determines exactly and instantaneously, at each instant of time, the active jump mode. However, such a device can be costly or it may not be even feasible. In principle, optimal control in the situation with no mode observation can be dealt with the theory of dynamic programming with imperfect state information, however, for the problem we are dealing with, this would lead to a nonlinear and high-dimensional optimization problem involving the information vector, also called the information state [30, 31]. Thus, it is reasonable to consider a control policy that is not a function of the active jump mode, and minimizes a suitable quadratic performance index. This is the scenario under investigation in this book.

The setup of MJLS with no mode observation is studied in the papers [5, 16, 22], and in the monograph [7, Sect. 3.5.2]. Notice that all of these approaches do not consider additive noise input, i.e., it is assumed that $w_k \equiv 0$. The paper [5] presents a

© The Author(s) 2016
A.N. Vargas et al., *Advances in the Control of Markov Jump Linear Systems with No Mode Observation*, SpringerBriefs in Control, Automation and Robotics, DOI 10.1007/978-3-319-39835-8_1

necessary optimal condition for the control problem in the receding horizon context with no noise; the papers [16, 22] deal with the H_2 control problem but the techniques based on LMI assure a guaranteed cost only; the monograph [7, Sect. 3.5.2] considers the stabilization problem taking the MJLS with no noise. Our results expand the knowledge of MJLS with additive noise input, and in this noisy setup, a method to compute the necessary optimal condition for the corresponding control problem is obtained.

The MJLS considered in this book is as follows. Let $(\Omega, \mathscr{F}, \{\mathscr{F}_k\}, P)$ be a fixed filtered probability space, and consider the discrete-time system

$$x_{k+1} = A_{\theta_k} x_k + B_{\theta_k} u_k + H_{\theta_k} w_k, \quad \forall k \geq 0, \ x_0 \in \mathscr{R}^r, \ \theta_0 \sim \pi_0, \tag{1}$$

where x_k, u_k, and w_k represent processes taking values in appropriately defined vector spaces. The noisy input $\{w_k\}$ forms an i.i.d. process with zero mean and stationary covariance matrix equals to the identity; and the process $\{\theta_k\}$ represents a discrete-time homogeneous Markov chain and takes values in the set $\mathscr{S} = \{1, \ldots, \sigma\}$.

The state of the system is formed by the pair (x_k, θ_k), and u_k denotes the control variable. The matrices $A_{\theta_k}, B_{\theta_k}$, and $H_{\theta_k}, k \geq 0$, have compatible dimensions, and for each $k \geq 0$,

$$(A_{\theta_k}, B_{\theta_k}, H_{\theta_k}) \in \{(A_1, B_1, H_1), \ldots, (A_\sigma, B_\sigma, H_\sigma)\}.$$

Figure 1 illustrates the working scheme of the jumps over the system. Note that the Markov chain θ_k drives the value of the system state x_{k+1} for all $k \geq 0$, according to (1).

Designing the control input u_k plays an important role in the theory of MJLS. In the MJLS literature, most of the results assume that the controller has complete and instantaneous access to the Markov state, but this assumption can fail in many real-time applications because the task of monitoring the Markovian mode requires a built-in sensor or a similar measurement instrument that might be expensive and difficult or even impossible to implement. In this case, a reasonable strategy is to use controllers whose implementation is irrespective of the Markov state. The design of optimal control for systems that do not have access to the Markovian mode is the central theme of this book.

Fig. 1 Jumping scheme: at each instant $k \geq 0$, just only one mode $\theta_k = i \in \mathscr{S}$ remains actived

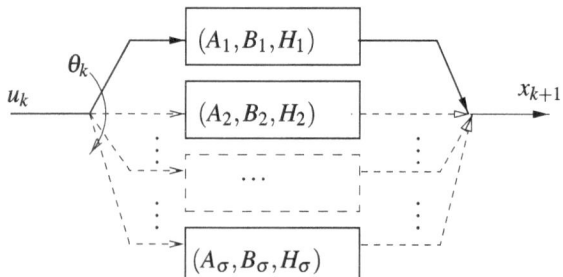

Seeking for simplicity and aiming at practical control applications, we assume that the controller has linear state feedback format with no mode observation, as in [7, p. 59], [5]:

$$u_k = G(k)x_k, \quad k \geq 0. \tag{2}$$

Note in (2) that the gain matrix $G(k)$ does not depend on the Markov state θ_k or on the conditional distribution $P(\theta_k | x_0, \ldots, x_k)$, $k \geq 0$, partly because the conditional distribution leads to a nonlinear filter that can be hard to implement, and partly because we seek for a sequence of gains $G(k)$ that can be precomputed offline.

Here, we are interested in evaluating the long-run behavior of the MJLS in (1); namely, for any given control input sequence $(u_0, u_1, \ldots)$, we introduce the long-run average cost

$$J(u_0, u_1, \ldots) = \limsup_{N \to \infty} \frac{1}{N} \sum_{k=0}^{N-1} \mathrm{E}_{x_0, \pi_0}[x_k' Q_{\theta_k} x_k + u_k' R_{\theta_k} u_k], \tag{3}$$

where $\mathrm{E}_{x_0, \pi_0}[\cdot] \equiv \mathrm{E}[\cdot | x_0, \pi_0]$ represents the expected value operator, and Q_{θ_k} and R_{θ_k} are positive semidefinite, symmetric matrices.

The control problem under study in this monograph is defined as follows.

$$J^* = \min_{u_0, u_1, \ldots} J(u_0, u_1, \ldots) \quad \text{s.t.} \quad (1) \text{ and } (2). \tag{4}$$

To the best of the authors' knowledge, the problem in (4) remains opened. Actually, this monograph presents an algorithm that attempts to solve the problem in (4).

The proposed algorithm assuredly computes a control that satisfies a necessary optimality condition for (4). This finding signifies that the control attains a local minimum, which could differ from the global minimum. How to attain the global minimum remains an open topic for research.

The main contribution of this monograph is twofold. First, it presents a numerical method that computes the necessary optimal condition for the control problem posed in (4). The method is based on a monotone strategy, iterated at each step by a variational approach, that produces the convergence to a set of gain matrices $\mathbf{G} = \{G(0), \ldots, G(N)\}$, $N > 0$, that satisfies the optimality condition.

The second contribution is the application of the gain sequence $\mathbf{G}$ to control the speed of a real DC motor device subject to abrupt power failures. The laboratory device is adapted to suffer power failures according to a homogeneous Markov chain. These elements constitute the main novelty of Chap. 2.

In Chap. 3, we deal with the approximating control problem. The idea behind this method is as follows. Suppose that $\mathbf{f} = \{f_0, \ldots, f_k, \ldots\}$ is a policy (control function). (see [32, 33]). If $\mathbf{F}$ denotes the set of all feasible policies $\mathbf{f}$, then the associated problem of N stages is defined as

$$J_N^* = \min_{\mathbf{f} \in \mathbf{F}} J_N.$$

The long-run average cost is defined as

$$\overline{J} = \limsup_{N \to \infty} \frac{J_N}{N}.$$

and the corresponding average cost problem is

$$\overline{J}^* = \min_{\mathbf{f} \in \mathbf{F}} \overline{J}.$$

Chapter 3 then contributes by showing the conditions under which the approximation

$$J_N^*/N \to \overline{J}^* \quad \text{as} \quad N \to \infty$$

holds true.

The results of Chaps. 2 and 3 are combined so as to solve the control problem (4).

References

1. M. Khanbaghi, R.P. Malhame, M. Perrier, Optimal white water and broke recirculation policies in paper mills via jump linear quadratic control. IEEE Trans. Automat. Control **10**(4), 578–588 (2002)
2. A.A.G. Siqueira, M.H. Terra, A fault tolerant manipulator robot based on H2, H-infinity, and mixed H2/H-infinity Markovian controls. IEEE/ASME Trans. Mechatron. **14**(2), 257–263 (2009)
3. A.N. Vargas, W. Furloni, J.B.R. do Val, Second moment constraints and the control problem of Markov jump linear systems. Numer. Linear Algebra Appl. **20**(2), 357–368 (2013)
4. O.L.V. Costa, W.L. de Paulo, Indefinite quadratic with linear costs optimal control of Markov jump with multiplicative noise systems. Automatica **43**, 587–597 (2007)
5. J.B.R. do Val, T. Başar, Receding horizon control of jump linear systems and a macroeconomic policy problem. J. Econ. Dyn. Control, **23**, 1099–1131 (1999)
6. J.P. Hespanha, P. Naghshtabrizi, Y. Xu, A survey of recent results in networked control systems. Proc. IEEE Spec. Issue Technol. Netw. Control Syst. **95**(1), 138–162 (2007)
7. O.L.V. Costa, M.D. Fragoso, R.P. Marques, *Discrete-Time Markovian Jump Linear Systems* (Springer, New York, 2005)
8. O.L.V. Costa, M.D. Fragoso, M.G. Todorov, *Continuous-Time Markov Jump Linear Systems*, Series: Probability and Its Applications (Springer, New York, 2013)
9. M. Mariton, *Jump Linear Systems in Automatic Control* (Marcel Dekker, New York, 1990)
10. P. Bolzern, P. Colaneri, G. De Nicolao, On almost sure stability of continuous-time Markov jump linear systems. Automatica **42**(6), 983–988 (2006)
11. P. Bolzern, P. Colaneri, G. De Nicolao, Almost sure stability of Markov jump linear systems with deterministic switching. IEEE Trans. Autom. Control **58**(1), 209–214 (2013)

12. Paolo Bolzern, Patrizio Colaneri, Giuseppe De Nicolao, Markov jump linear systems with switching transition rates: mean square stability with dwell-time. Automatica **46**(6), 1081–1088 (2010)
13. E.F. Costa, J.B.R. do Val, On the detectability and observability of continuous-time Markov jump linear systems. SIAM J. Control Optim. **41**(4), 1295–1314 (2002)
14. E.F. Costa, J.B.R. do Val, M.D. Fragoso, A new approach to detectability of discrete-time Markov jump linear systems. SIAM J. Control Optim. **43**(6), 2132–2156 (2005)
15. E.F. Costa, A.N. Vargas, J.B.R. do Val, Quadratic costs and second moments of jump linear systems with general Markov chain. Math. Control Signals Syst. **23**(1), 141–157 (2011)
16. J.B.R. do Val, J.C. Geromel, A.P. Gonçalves, The H_2 control for jump linear systems: cluster observations of the Markov state. Automatica **38**, 343–349 (2002)
17. V. Dragan, T. Morozan, Exponential stability in mean square for a general class of discrete-time linear stochastic systems. Stoch. Anal. Appl. **26**(3), 495–525 (2008)
18. F. Dufour, R.J. Elliott, Adaptive control of linear systems with Markov perturbations. IEEE Trans. Autom. Control **43**(3), 351–372 (1998)
19. J.C. Geromel, A.P.C. Gonçalves, A.R. Fioravanti, Dynamic output feedback control of discrete-time Markov jump linear systems through linear matrix inequalities. SIAM J. Control Optim. **48**(2), 573–593 (2009)
20. Fanbiao Li, Wu Ligang, Peng Shi, Cheng-Chew Lim, State estimation and sliding mode control for semi-Markovian jump systems with mismatched uncertainties. Automatica **51**, 385–393 (2015)
21. N. Meskin, K. Khorasani, A geometric approach to fault detection and isolation of continuous-time Markovian jump linear systems. IEEE Trans. Automat. Control **55**(6), 1343–1357 (2010)
22. R.C.L.F. Oliveira, A.N. Vargas, J.B.R. do Val, P.L.D. Peres, Robust stability, H2 analysis and stabilisation of discrete-time Markov jump linear systems with uncertain probability matrix. Int. J. Control, **82**(3) 470–481 (2009)
23. Peng Shi, Fanbiao Li, A survey on Markovian jump systems: modeling and design. Int. J. Control Autom. Syst. **13**(1), 1–16 (2015)
24. Peng Shi, Yanyan Yin, Fei Liu, Jianhua Zhang, Robust control on saturated Markov jump systems with missing information. Inf. Sci. **265**, 123–138 (2014)
25. M. Tanelli, B. Picasso, P. Bolzern, P. Colaneri, Almost sure stabilization of uncertain continuous-time Markov jump linear systems. IEEE Trans. Autom. Control **55**(1), 195–201 (2010)
26. M.G. Todorov, M.D. Fragoso, On the robust stability, stabilization, and stability radii of continuous-time infinite Markov jump linear systems. SIAM J. Control Optim. **49**, 1171–1196 (2011)
27. J.B.R. Do Val, E.F. Costa, Stabilizability and positiveness of solutions of the jump linear quadratic problem and the coupled algebraic Riccati equation. IEEE Trans. Automat. Control **50**(5), 691–695 (2005)
28. O.L.V. Costa, E.F. Tuesta, Finite horizon quadratic optimal control and a separation principle for Markovian jump linear systems. IEEE Trans. Automat. Control **48**, (2003)
29. M.D. Fragoso, O.L.V. Costa, A separation principle for the continuous-time LQ-problem with Markovian jump parameters. IEEE Trans. Automat. Control **55**(12), 2692–2707 (2010)
30. D.P. Bertsekas, *Dynamic Programming and Optimal Control* (Athena Scientific, Nashua, 2007)
31. P.R. Kumar, P. Varaiya, *Stochastic Systems: Estimation, Identification and Adaptive Control* (Prentice-Hall Inc, New Jersey, 1986)
32. D.P. Bertsekas, S.E. Shreve, *Stochastic Optimal Control: The Discrete Time Case* (Athena Scientific, Belmont, 1996)
33. O. Hernández-Lerma, J.B. Lasserre, *Discrete-Time Markov Control Processes: Basic Optimality Criteria* (Springer, New York, 1996)

Finite-Time Control Problem

1 Finite-Time Control Problem: Variational Method

Recall the Markov jump linear system (see (1))

$$x_{k+1} = A_{\theta_k} x_k + B_{\theta_k} u_k + H_{\theta_k} w_k, \quad \forall k \geq 0, \ x_0 \in \mathscr{R}^r, \ \theta_0 \sim \pi_0,$$

where x_k, u_k, and w_k represent the state, control input, and noisy disturbances, respectively.

Consider the standard Nth horizon quadratic cost

$$J_N(x_0, \pi_0) := \mathrm{E}_{x_0, \pi_0} \left[\sum_{k=0}^{N-1} (x_k' Q_{\theta_k} x_k + u_k' R_{\theta_k} u_k) + x_N' F_{\theta_N} x_N \right], \tag{1}$$

In view of the identity in (7), we have

$$J_N(x_0, \theta_0) = \langle L(0), X(0) \rangle + \pi(0)' \omega(0). \tag{2}$$

Thus, the stochastic control problem posed in (36) can be recast as that of minimizing the deterministic functional in (2) with respect to the gain sequence $\mathbf{G} \in \mathscr{G}$. This fact lead us to focus the remaining analysis on the deterministic expression of (2).

2 Definitions and Basic Concepts

Let $\mathscr{R}^r$ denote the usual rth dimensional Euclidean space, and let $\mathscr{M}^{r,s}$ ($\mathscr{M}^r$) represent the linear space formed by all $r \times s$ ($r \times r$) real matrices. Let $\mathscr{S}^r$ represent the normed linear subspace of $\mathscr{M}^r$ of symmetric matrices such as $\{U \in \mathscr{M}^r : U = U'\}$, where U' denotes the transpose of U. Consider also $\mathscr{S}^{r0}$ ($\mathscr{S}^{r+}$) its closed (open)

© The Author(s) 2016
A.N. Vargas et al., *Advances in the Control of Markov Jump Linear Systems with No Mode Observation*, SpringerBriefs in Control, Automation and Robotics, DOI 10.1007/978-3-319-39835-8_2

convex cone of positive semidefinite (definite) matrices $\{U \in \mathscr{S}^r : U \geq 0 \ (> 0)\}$. Let $\mathscr{S} := \{1, \ldots, \sigma\}$ be a finite set, and let $\mathbb{M}^{r,s}$ denote the linear space formed by a number σ of matrices such that $\mathbb{M}^{r,s} = \{U = (U_1, \ldots, U_\sigma) : U_i \in \mathscr{M}^{r,s}, i \in \mathscr{S}\}$; also $\mathbb{M}^r \equiv \mathbb{M}^{r,r}$. Moreover, we set $\mathbb{S}^r = \{U = (U_1, \ldots, U_\sigma) : U_i \in \mathscr{S}^r, i \in \mathscr{S}\}$, and we write $\mathbb{S}^{r0}$ ($\mathbb{S}^{r+}$) when $U_i \in \mathscr{S}^{r0}$ ($\in \mathscr{S}^{r+}$) for all $i \in \mathscr{S}$.

We employ the ordering $U > V$ ($U \geq V$) for elements of $\mathbb{S}^r$, meaning that $U_i - V_i$ is positive definite (semi-definite) for all $i \in \mathscr{S}$, and similarly for other mathematical relations. Let $\mathrm{tr}\{\cdot\}$ be the trace operator. When applied for some $U \in \mathbb{S}^n$, the operator $\mathrm{tr}\{U\}$ signifies $(\mathrm{tr}\{U_1\}, \ldots, \mathrm{tr}\{U_\sigma\})$. Define the inner product on the space $\mathbb{M}^{r,s}$ as

$$\langle U, V \rangle = \sum_{i=1}^{\sigma} \mathrm{tr}\{U_i' V_i\}, \quad \forall V, U \in \mathbb{M}^{r,s},$$

and the Frobenius norm $\|U\|_2^2 = \langle U, U \rangle$.

The transition probability matrix is denoted by $\mathbb{P} = [p_{ij}]$, for all $i, j \in \mathscr{S}$. The state of the Markov chain at a certain time k is determined according to an associated probability distribution $\pi(k)$ on $\mathscr{S}$, namely, $\pi_i(k) := \Pr(\theta_k = i)$. Considering the column vector $\pi(k) = [\pi_1(k), \ldots, \pi_\sigma(k)]'$, the state distribution of the chain, $\pi(k)$, is defined as $\pi(k) = (\mathbb{P}')^k \pi(0)$. Given $U \in \mathbb{M}^{r,s}$ and $\pi(k) \in \mathscr{R}^\sigma$, $k \geq 0$, we let $\pi(k)U$ represent the operation $(\pi_1(k)U_1, \ldots, \pi_\sigma(k)U_\sigma)$.

Associated with the system (1)–(36), we define $A \in \mathbb{M}^r, B \in \mathbb{M}^{r,s}, H \in \mathbb{M}^{r,q}, Q \in \mathbb{S}^{r0}$, and $R \in \mathbb{S}^{s+}$. In addition, we define the operators $\mathscr{D} = \{\mathscr{D}_i, i \in \mathscr{S}\} : \mathbb{S}^{n0} \mapsto \mathbb{S}^{n0}$ and $\mathscr{E} = \{\mathscr{E}_i, i \in \mathscr{S}\} : \mathbb{S}^{n0} \mapsto \mathbb{S}^{n0}$, respectively, as

$$\mathscr{D}_i(U) := \sum_{j=1}^{\sigma} p_{ji} U_j, \quad \mathscr{E}_i(U) := \sum_{j=1}^{\sigma} p_{ij} U_j, \quad \forall i \in \mathscr{S}, \ \forall U \in \mathbb{S}^{n0}. \qquad (3)$$

The class of all admissible gain sequences $\mathbf{G} = \{G(0), \ldots, G(N-1)\}$ as in (2) is represented by $\mathscr{G}$. Note that the corresponding closed-loop matrix sequence $A(k) \in \mathbb{M}^r$ satisfies

$$A_i(k) := A_i + B_i G(k), \quad \forall i \in \mathscr{S}, \ k = 0, \ldots, N-1.$$

Let us define the conditional second moment matrix of the system state x_k, $k \geq 0$, as

$$X_i(k) = \mathrm{E}[x_k x_k' \, \mathrm{1\!l}_{\{\theta_k = i\}}], \quad \forall i \in \mathscr{S}, \ \forall k \geq 0,$$

where $\mathrm{1\!l}_{\{\cdot\}}$ stands for the Dirac measure. Setting $X(k) = \{X_1(k), \ldots, X_\sigma(k)\} \in \mathbb{S}^{n0}$ for every $k \geq 0$, we obtain the recurrence [1, Proposition 3.35]

$$X(k+1) = \mathscr{D}\big(A(k)X(k)A(k)' + \pi(k)HH'\big), \quad \forall k \geq 0, \qquad (4)$$

with $X_i(0) = \pi_i(0)x_0 x_0'$ for each $i \in \mathcal{S}$. In addition, let us define the sets $L(k) \in \mathbb{S}^{r0}$ and $\omega(k) \in \mathbb{S}^{10}$, $k = 0, \ldots, N$, from the coupled recurrence equations

$$L_i(k) = Q_i + G(k)'R_i G(k) + A_i(k)'\mathcal{E}_i(L(k+1))A_i(k), \quad L_i(N) = F_i, \quad \forall i \in \mathcal{S} \quad (5)$$

and

$$\omega_i(k) = \mathcal{E}_i(\omega(k+1)) + \text{tr}\{\mathcal{E}_i(L(k+1))H_i H_i'\}, \quad \omega_i(N) = 0, \quad \forall i \in \mathcal{S}. \quad (6)$$

Lemma 2.1 *Given* $\mathbf{G} \in \mathcal{G}$, *the next set of identities hold for each* $k = 0, \ldots, N-1$:

$$\mathrm{E}_{x_0, \pi_0}\left[\sum_{\ell=k}^{N-1} x_\ell'(Q_{\theta_\ell} + G(\ell)'R_{\theta_\ell}G(\ell))x_\ell + x_N' F_{\theta_N} x_N \right]$$

$$= \sum_{\ell=k}^{N-1} \sum_{i=1}^{\sigma} \text{tr}\{(Q_i + G(\ell)'R_i G(\ell))X_i(\ell) + F_i X_i(N)\}$$

$$= \langle L(k), X(k) \rangle + \pi(k)'\omega(k). \quad (7)$$

Proof Given $\mathbf{G} \in \mathcal{G}$, let us define the random variable $W(t, \cdot)$ for each $t = 0, \ldots, N$ as

$$W(t, x_t, \theta_t) = \mathrm{E}\left[\sum_{\ell=t}^{N-1} x_\ell'(Q_{\theta_\ell} + G(\ell)'R_{\theta_\ell}G(\ell))x_\ell + x_N' F_{\theta_N} x_N \,\Big|\, x_t, \theta_t \right], \quad (8)$$

with terminal condition $W(N, x_N, \theta_N) = x_N' F_{\theta_N} x_N$. Since the joint process $\{x_t, \theta_t\}$ is Markovian [1, p. 31], we can write the identity

$$W(t, x_t, \theta_t) = x_t'(Q_{\theta_t} + G(t)'R_{\theta_t}G(t))x_t$$

$$+ \mathrm{E}\left[\mathrm{E}\left[\sum_{\ell=t+1}^{N-1} x_\ell'(Q_{\theta_\ell} + G(\ell)'R_{\theta_\ell}G(\ell))x_\ell \right.\right.$$

$$\left.\left. + x_N' F_{\theta_N} x_N \,\Big|\, x_{t+1}, \theta_{t+1} \right] \,\Big|\, x_t, \theta_t \right]$$

$$= x_t'(Q_{\theta_t} + G(t)'R_{\theta_t}G(t))x_t + \mathrm{E}[W(t+1, x_{t+1}, \theta_{t+1}) \mid x_t, \theta_t]. \quad (9)$$

Setting $x_t = x \in \mathcal{R}^r$ and $\theta_t = i \in \mathcal{S}$, we now show by induction that

$$W(t, x, i) = x'L_i(t)x + \omega_i(t), \quad (10)$$

where $L(t) \in \mathbb{S}^{r0}$ and $\omega(t) \in \mathbb{M}^1$, $t = 0, \ldots, N$, satisfy (5) and (6), respectively. Indeed, take $t = N$ and it is immediate that

$$W(N, x_N, \theta_N) = x_N' F_{\theta_N} x_N = x_N' L_{\theta_N}(N)x_N,$$

which shows the result for $t = N$. Now, suppose that (10) holds for $t = m + 1$, i.e., that

$$W(m + 1, x_{m+1}, \theta_{m+1}) = x'_{m+1} L_{\theta_{m+1}}(m + 1)x_{m+1} + \omega_{\theta_{m+1}}(m + 1)$$

is valid. We then get from (9) that

$$\begin{aligned} W(m, x, i) = {} & x'(Q_i + G(m)'R_i G(m))x \\ & + \mathrm{E}\big[x'_{m+1} L_{\theta_{m+1}}(m + 1)x_{m+1} + \omega_{\theta_{m+1}}(m + 1) \mid \theta_m = i, x_m = x\big]. \end{aligned}$$

The right-hand side of the above identity is equal to

$$\begin{aligned} x'\Big(Q_i + G(m)'R_i G(m) + A_i(m)'\mathscr{E}_i(L(m + 1))A_i(m)\Big)x + \mathscr{E}_i(\omega(m + 1)) \\ + \mathrm{E}\big[\mathrm{tr}\{L_{\theta_{m+1}}(m + 1)H_i w(m)w(m)'H'_i\} \mid \theta_m = i, x_m = x\big]. \end{aligned}$$

Since the last term in this expression is equal to $\mathrm{tr}\{\mathscr{E}_i(L(m + 1))H_i H'_i\}$, we can conclude that

$$\begin{aligned} W(m, x, i) = {} & x'\Big(Q_i + G(m)'R_i G(m) + A_i(m)'\mathscr{E}_i(L(m + 1))A_i(m)\Big)x \\ & + \mathscr{E}_i(\omega(m + 1)) + \mathrm{tr}\{\mathscr{E}_i(L(m + 1))H_i H'_i\} \\ = {} & x'L_i(m)x + \omega_i(m), \end{aligned}$$

which shows the result in (10) for $t = m$. This induction argument completes the proof of (10). The result of Lemma 2.1 then follows from the conditional expectation property together with (8) and (10). $\qquad\square$

2.1 Main Results

For sake of clarity, let us represent by $J_\mathbf{G}$ the cost $J_N(x_0, \theta_0)$ when evaluated for $\mathbf{G} \in \mathscr{G}$. The next result presents the necessary optimal condition for the considered control problem.

Theorem 2.1 (Necessary optimal condition) *Suppose that $\mathbf{G} = \{G(0), \ldots, G(N - 1)\} \in \mathscr{G}$ is such that $J_\mathbf{G} = \min_{\mathbf{K} \in \mathscr{G}} J_\mathbf{K}$. Then, for each $k = 0, \ldots, N - 1$,*

$$\sum_{i=1}^{\sigma} [(R_i + B'_i \mathscr{E}_i(L(k + 1))B_i)G(k) + B'_i \mathscr{E}_i(L(k + 1))A_i]X_i(k) = 0, \qquad (11)$$

where $X(k) \in \mathbb{S}^{r_0}$ and $L(k) \in \mathbb{S}^{r_0}$ are as in (4) and (5), respectively.

Proof For some $k \geq 0$, let us assume that $G(0), \ldots, G(k-1), G(k+1), \ldots,$ $G(N-1)$ are fixed optimal minimizers and $G(k)$ is a free design variable. In this case, both $X(0), \ldots, X(k)$ and $L(k+1), \ldots, L(N)$ are fixed, and this allows us to deduce from the identity of (7) that

$$\arg\min_{G(k)} J_{\mathbf{G}} = \arg\min_{G(k)} \langle L(k), X(k) \rangle + \pi(k)'\omega(k).$$

Hence,

$$\arg\min_{G(k)} J_{\mathbf{G}} = \arg\min_{G(k)} \left[\sum_{i=1}^{\sigma} \mathrm{tr}\{[Q_i + G(k)'R_iG(k) \right.$$
$$+ (A_i + B_iG(k))'\mathscr{E}_i(L(k+1))(A_i + B_iG(k))]X_i(k)\}$$
$$\left. + \sum_{i=1}^{\sigma} \pi_i(k) \left(\mathscr{E}_i(\omega(k+1)) + \mathrm{tr}\{\mathscr{E}_i(L(k+1))H_iH_i'\} \right) \right].$$

Taking the differentiation with respect to $G(k)$ in the expression within the last brackets, we obtain the expression in (11) and the proof is completed.   □

Remark 2.1 An interesting open question is whether the necessary optimal condition of Theorem 2.1 is also sufficient. Convexity can not be used to conclude sufficiency because the optimization approach is not convex at all, as the next example illustrates.

Example 2.1 Consider the single-input single-output MJLS as in (1) with parameters $A_1 = 0.3$, $A_2 = 0.1$, $B_1 = -1$, $B_2 = 1$, $H_i = 0$, $Q_i = 0.4$, $R_i = 1$, $F_i = 0.5$, $i = 1, 2, N = 2, x(0) = 2$, and $\pi_0 = [0.25 \ 0.75]$. We consider the stochastic matrix $\mathbb{P} = [p_{ij}]$, $i, j = 1, 2$ as $p_{11} = 0.6$, $p_{12} = 0.4$, $p_{12} = 0.2$, and $p_{22} = 0.8$. After some algebraic manipulations on (2), one can rewrite the cost equivalently as

$$J_{\{G(0),G(1)\}} = 1.6 + 4G(0)^2 + (0.4 + G(1)^2)(0.3 - G(0))^2 + (1.2 + 3G(1)^2)(0.1 + G(0))^2$$
$$+ 0.3(0.3 - G(1))^2(0.3 - G(0))^2 + 0.2(0.1 + G(1))^2(0.3 - G(0))^2$$
$$+ 0.1(0.3 - G(1))^2(0.1 + G(0))^2 + 0.4(0.1 + G(1))^2(0.1 + G(0))^2. \quad (12)$$

The functional in (12) is not convex as one can inspect in the contour plot of Fig. 1. Note also in the figure that the function has a unique minimum with multiple solutions.

Remark 2.2 It should be noted that the coupled equations (4), (5), and (11) are nonlinear with respect to $\mathbf{G} = \{G(0), \ldots, G(N-1)\} \in \mathscr{G}$, and their evaluation represents a challenge for analytical and numerical fronts. The method of the next section represents a contribution towards this direction since it computes $\mathbf{G} \in \mathscr{G}$ that satisfies simultaneously (4), (5), and (11).

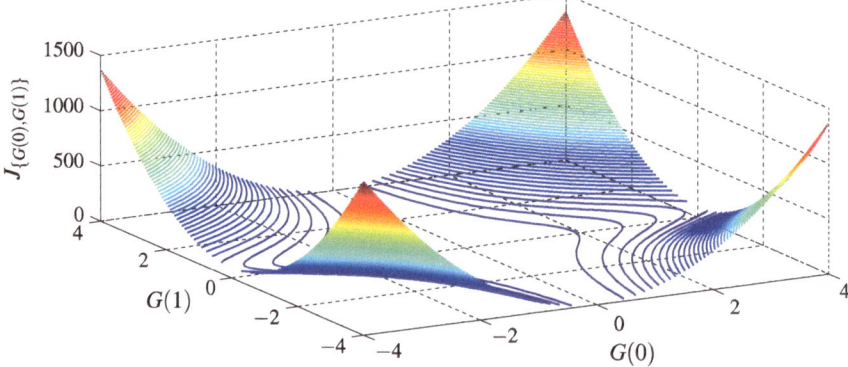

Fig. 1 Three-dimensional contour plot of the cost $J_{\{G(0),G(1)\}}$ as in Example 2.1

2.2 Numerical Method for the Necessary Optimal Condition

In this section, we provide a method for evaluating the necessary optimal condition of Theorem 2.1. The idea of the method is to employ a variational principle to produce monotone cost functions. On the convergence, it provides a gain sequence $\mathbf{G} \in \mathscr{G}$ that satisfies the optimality condition of Theorem 2.1.

To present the method, we require some additional notation. Let $\eta = 0, 1, \ldots$ be an iteration index. For some given sequence

$$\mathbf{G}[\eta] := \{G^{[\eta]}(0), \ldots, G^{[\eta]}(N-1)\} \in \mathscr{G}, \ \eta = 0, 1, \ldots.$$

let us define

$$A_i^{[\eta]}(k) := A_i + B_i G^{[\eta]}(k), \quad Q_i^{[\eta]}(k) := Q_i + G^{[\eta]}(k)' R_i G^{[\eta]}(k),$$
$$\forall i \in \mathscr{S}, \ k = 0, \ldots, N-1.$$

Let us now consider the following algorithm.

- *Step 1*: Set the iterations counter $\eta = 0$. Pick an arbitrary initial sequence $\mathbf{G}[0] \in \mathscr{G}$.
- *Step 2*: For each $k = 1, \ldots, N$, find $X^{[\eta]}(k) \in \mathbb{S}^{r0}$, the solution of the following set of equations:

$$X^{[\eta]}(k) = \mathscr{D}\Big(A^{[\eta]}(k-1)X^{[\eta]}(k-1)A^{[\eta]}(k-1)' + \pi(k-1)HH'\Big),$$

with $X^{[\eta]}(0) = X(0)$. Set $\eta = \eta + 1$ and go to *Step 3*.
- *Step 3*: Set $L^{[\eta]}(N) = F$ and $k = N - 1$. Let $G^{[\eta]}(k) \in \mathscr{M}^{s,r}$ be defined by

$$\sum_{i=1}^{\eta}\left[\left(R_i + B_i'\mathscr{E}_i\left(L^{[\eta]}(k+1)\right)B_i\right)G^{[\eta]}(k) + B_i'\mathscr{E}_i\left(L^{[\eta]}(k+1)\right)A_i\right]X_i^{[\eta-1]}(k) = 0.$$

(13)

Evaluate the expression

$$(G^{[\eta]}(k) - G^{[\eta-1]}(k))X_i^{[\eta-1]}(k) = 0$$

(14)

and set $G^{[\eta]}(k) = G^{[\eta-1]}(k)$ only if (14) holds true for all $i \in \mathscr{S}$. Compute $L^{[\eta]}(k) \in \mathbb{S}^{r0}$ and $\omega^{[\eta]}(k) \in \mathbb{S}^1$ from the recurrence

$$L^{[\eta]}(k) = Q^{[\eta]}(k) + A^{[\eta]}(k)'\mathscr{E}(L^{[\eta]}(k+1))A^{[\eta]}(k), \quad L^{[\eta]}(N) = F,$$

(15)

$$\omega^{[\eta]}(k) = \mathscr{E}(\omega^{[\eta]}(k+1)) + \mathrm{tr}\{\mathscr{E}(L^{[\eta]}(k+1))HH'\}, \quad \omega^{[\eta]}(N) = 0.$$

(16)

Set $k = k - 1$; if $k \geq 0$ then return to the beginning of *Step 3*.
- *Step 4*: Compute the cost $J_{\mathbf{G}[\eta]} = \langle L^{[\eta]}(0), X(0)\rangle + \pi(0)'\omega^{[\eta]}(0)$. If the evaluation of the difference $J_{\mathbf{G}[\eta-1]} - J_{\mathbf{G}[\eta]}$ is sufficiently small, then stop the algorithm. Otherwise, return to the beginning of *Step 2*.

Remark 2.3 The equation (13) can be transformed into a system of linear equations of the form $\mathscr{A}w = b$, whose solution can be obtained to a desired precision using efficient numerical methods available in literature. In fact, by applying the Kronecker product and the column stacking operator (denoted by $\otimes$ and vec, respectively) one obtains $w = \mathrm{vec}\left(G^{[\eta]}(k)\right)$,

$$\mathscr{A} = \sum_{i=1}^{\eta}\left[X_i^{[\eta-1]}(k) \otimes \left(R_i + B_i'\mathscr{E}_i\left(L^{[\eta]}(k+1)\right)B_i\right)\right],$$

and

$$b = -\mathrm{vec}\left[\sum_{i=1}^{\eta}\left[B_i'\mathscr{E}_i\left(L^{[\eta]}(k+1)\right)A_i\right]X_i^{[\eta-1]}(k)\right].$$

Remark 2.4 The algorithm can be implemented in a receding horizon framework. At each time instant $\ell \geq 0$, the gain sequence $G(t)$, $\ell \leq t \leq \ell + N - 1$ is calculated and only $G(t)$, $t = \ell$ is implemented. In order to compute this gain sequence, assuming $X(\ell) = M$ with M given, one can employ the algorithm with a time displacement $k = t - \ell$: set $X(0) = M$ and obtain the gain sequence $G(k)$, $0 \leq k \leq N - 1$ as indicated in the algorithm (producing $G(t)$, $\ell \leq t \leq \ell + N - 1$). Note that only the covariance matrix $X(\ell) = M$ (or an estimate of it) is required at time instant ℓ to obtain the corresponding receding horizon gain.

Theorem 2.2 *The gain sequences* $\mathbf{G}[\eta] \in \mathscr{G}$, $\eta = 0, 1, \ldots$, *generated in the algorithm of Steps 1–4, satisfy the monotone property* $J_{\mathbf{G}[\eta]} \geq J_{\mathbf{G}[\eta+1]}$. *Moreover, the limit* $\mathbf{G} = \lim_{\eta\to\infty} \mathbf{G}[\eta]$ *exists and it satisfies the necessary optimal condition of Theorem 2.1.*

Proof We divide the proof of Theorem 2.2 into two parts. The first part introduces an evaluation for the cost corresponding to two different gain sequences, and the second one proves that the cost corresponding to the gain sequences from *Step 3* is monotonically non-increasing. As a byproduct, we get that the gain sequences converge to a sequence that satisfies the optimality condition of Theorem 2.1.

To begin with, we need to introduce some additional notation. For a given gain sequence $\mathbf{G} = \{G(0), \ldots, G(N-1)\} \in \mathscr{G}$, let us consider the operator

$$\mathscr{L}_{i,\mathbf{G}}^k(U) := (A_i + B_iG(k))'\mathscr{E}_i(U)(A_i + B_iG(k)),$$
$$k = 0, \ldots, N-1, \ \forall i \in \mathscr{S}, \ U \in \mathbb{S}^r,$$

so that we can write

$$L_{i,\mathbf{G}}(k) = Q_i + G(k)'R_iG(k) + \mathscr{L}_{i,\mathbf{G}}^k(L_{\mathbf{G}}(k+1)), \quad k = 0, \ldots, N-1, \ \forall i \in \mathscr{S},$$

with $L_{\mathbf{G}}(N) = F$.

After some algebraic manipulation (see Appendix for a detailed proof), we have

$$L_{\mathbf{G}}(k) - L_{\mathbf{K}}(k) = \delta_{\mathbf{G},\mathbf{K}}^k + \mathscr{L}_{\mathbf{K}}^k(L_{\mathbf{G}}(k+1) - L_{\mathbf{K}}(k+1)), \quad k = 0, \ldots, N-1, \tag{17}$$

with both $\mathbf{G}$ and $\mathbf{K}$ belonging to $\mathscr{G}$, where

$$\delta_{i,\mathbf{G},\mathbf{K}}^k := (G(k) - Z_i^k)'\Lambda_{i,\mathbf{G}}^{k+1}(G(k) - Z_i^k) - (K(k) - Z_i^k)'\Lambda_{i,\mathbf{G}}^{k+1}(K(k) - Z_i^k), \ \forall i \in \mathscr{S}, \tag{18}$$

with $\Lambda_{i,\mathbf{G}}^k := R_i + B_i'\mathscr{E}_i(L_{\mathbf{G}}(k))B_i$ and $Z_i^k := -(\Lambda_{i,\mathbf{G}}^{k+1})^{-1}B_i'\mathscr{E}_i(L_{\mathbf{G}}(k+1))A_i$. Moreover, if $\mathbf{G} = \mathbf{G}[\eta] \in \mathscr{G}$ is the gain sequence that satisfies (13) and $X(k) = X^{[\eta-1]}(k), k = 0, \ldots, N$, is the corresponding second moment trajectory from *Step 2*, then we have [2, p. 1123]

$$\langle X(k), \delta_{\mathbf{G},\mathbf{K}}^k \rangle = \|(\Lambda_{\mathbf{G}}^{k+1})^{\frac{1}{2}}(G(k) - Z^k)X(k)^{\frac{1}{2}}\|_2^2 - \|(\Lambda_{\mathbf{G}}^{k+1})^{\frac{1}{2}}(K(k) - Z^k)X(k)^{\frac{1}{2}}\|_2^2$$
$$= -\|(\Lambda_{\mathbf{G}}^{k+1})^{\frac{1}{2}}(G(k) - K(k))X(k)^{\frac{1}{2}}\|_2^2. \tag{19}$$

The expression of (19) will be useful on evaluating the quantity $J_{\mathbf{G}[\eta]} - J_{\mathbf{G}[\eta-1]}$. Indeed, we derive in the sequel the arguments to show that

$$J_{\mathbf{G}[\eta]} - J_{\mathbf{G}[\eta-1]} = \langle X(k), \delta_{\mathbf{G}}^k \rangle = -\|(\Lambda_{\mathbf{G}}^{k+1})^{\frac{1}{2}}(G(k) - K(k))X(k)^{\frac{1}{2}}\|_2^2. \tag{20}$$

This result is important because it enables us to conclude that the cost sequence generated by *Steps 1–4* is monotone, i.e., there holds $J_{\mathbf{G}[\eta-1]} \geq J_{\mathbf{G}[\eta]}$ for every $\eta = 1, 2, \ldots$, thus showing the first statement of Theorem 2.2.

To show the identity in (20), let us define the sequences

$$\mathscr{G}^{[\eta,k]} := \{G^{[\eta-1]}(0), \ldots, G^{[\eta-1]}(k-1), G^{[\eta]}(k), \ldots, G^{[\eta]}(N)\}, \tag{21}$$

for each $k = 0 \ldots, N$, and set $\mathscr{G}^{[\eta]} = \mathscr{G}^{[\eta,0]}$. Recall the expression of the cost in (7), and note that the last element of $\mathscr{G}^{[\eta,k]}$, i.e., $G^{[\eta]}(N)$, does not influence the value of the cost, so that $J_{\mathscr{G}^{[\eta-1]}} = J_{\mathscr{G}^{[\eta,N]}}$.

Step 3 calculates $G^{[\eta]}(k)$ backwards in time. Thus, when an iteration of *Step 3* occurs, the element $G^{[\eta-1]}(k)$ in $\mathscr{G}^{[\eta,k+1]}$ is modified to $G^{[\eta]}(k)$ in $\mathscr{G}^{[\eta,k]}$, while the other elements remain unchanged. This observation leads to

$$m > k \implies L_{\mathscr{G}^{[\eta,k]}}(m) = L_{\mathscr{G}^{[\eta,k+1]}}(m). \tag{22}$$

Let us now define the recurrence

$$\omega_{i,G}(k) = \mathscr{E}_i(\omega_G(k+1)) + \mathrm{tr}\{\mathscr{E}_i(L_G(k+1))H_iH_i'\}, \quad \omega_{i,G}(N) = 0, \quad \forall i \in \mathscr{S}. \tag{23}$$

It follows from (23) that

$$\begin{aligned}
\omega_{\mathscr{G}^{[\eta,k]}}(k) - \omega_{\mathscr{G}^{[\eta,k+1]}}(k) &= \mathscr{E}\left(\omega_{\mathscr{G}^{[\eta,k]}}(k+1) - \omega_{\mathscr{G}^{[\eta,k+1]}}(k+1)\right) \\
&\quad + \mathrm{tr}\{\mathscr{E}\left(L_{\mathscr{G}^{[\eta,k]}}(k+1) - L_{\mathscr{G}^{[\eta,k+1]}}(k+1)\right)HH'\}
\end{aligned} \tag{24}$$

with $\omega_{\mathscr{G}^{[\eta,k]}}(N) = \omega_{\mathscr{G}^{[\eta,k+1]}}(N) = 0$. Since the rightmost term of (24) is null due to the identity in (22), we can apply a simple induction argument on the resulting expression from (24) to conclude that

$$\omega_{\mathscr{G}^{[\eta,k]}}(k) = \omega_{\mathscr{G}^{[\eta,k+1]}}(k), \quad k = 0, \ldots, N-1. \tag{25}$$

One can employ a similar reasoning for the recurrence in (4) to show that

$$k \geq m \geq 0 \implies X_{\mathscr{G}^{[\eta,k]}}(m) = X_{\mathscr{G}^{[\eta,k+1]}}(m). \tag{26}$$

In particular, we can observe from *Step 2* the validity of the identity

$$X_{\mathscr{G}^{[\eta,k]}}(k) = X^{[\eta-1]}(k), \quad k = 0, \ldots, N. \tag{27}$$

Now, we are able to prove the identity in (20). Indeed, from (7), we have

$$\begin{aligned}
J_{\mathscr{G}^{[\eta,k]}} - J_{\mathscr{G}^{[\eta,k+1]}} &= \left\langle X_{\mathscr{G}^{[\eta,k]}}(k), L_{\mathscr{G}^{[\eta,k]}}(k)\right\rangle + \pi(k)'\omega_{\mathscr{G}^{[\eta,k]}}(k) \\
&\quad - \left\langle X_{\mathscr{G}^{[\eta,k+1]}}(k), L_{\mathscr{G}^{[\eta,k+1]}}(k)\right\rangle - \pi(k)'\omega_{\mathscr{G}^{[\eta,k+1]}}(k).
\end{aligned} \tag{28}$$

Now substituting (25)–(27) into (28), we get

$$J_{\mathscr{G}^{[\eta,k]}} - J_{\mathscr{G}^{[\eta,k+1]}} = \left\langle X^{[\eta-1]}(k), L_{\mathscr{G}^{[\eta,k]}}(k) - L_{\mathscr{G}^{[\eta,k+1]}}(k)\right\rangle,$$

or equivalently, we can invoke the identity of (17) to obtain

$$J_{\mathscr{G}^{[\eta,k]}} - J_{\mathscr{G}^{[\eta,k+1]}}$$
$$= \big\langle X^{[\eta-1]}(k),\ \delta^k_{\mathscr{G}^{[\eta,k]},\mathscr{G}^{[\eta,k+1]}} + \mathscr{L}^k_{\mathscr{G}^{[\eta,k+1]}}\big(L_{\mathscr{G}^{[\eta,k]}}(k+1) - L_{\mathscr{G}^{[\eta,k+1]}}(k+1)\big)\big\rangle. \quad (29)$$

But then we can employ (22) with $m = k + 1$ to conclude that

$$J_{\mathscr{G}^{[\eta,k]}} - J_{\mathscr{G}^{[\eta,k+1]}} = \big\langle X^{[\eta-1]}(k),\ \delta^k_{\mathscr{G}^{[\eta,k]},\mathscr{G}^{[\eta,k+1]}}\big\rangle. \quad (30)$$

Hence, if we let

$$\xi(k) := \|(\Lambda^{k+1}_{\mathscr{G}^{[\eta,k]}})^{\frac{1}{2}}(G^{[\eta]}(k) - G^{[\eta-1]}(k))X^{[\eta-1]}(k)^{\frac{1}{2}}\|_2^2, \quad k = 0, \ldots, N,$$

then we can combine (19) and (30) to write

$$J_{\mathscr{G}^{[\eta,k]}} - J_{\mathscr{G}^{[\eta,k+1]}} = -\xi(k), \quad k = 0, \ldots, N.$$

Since the matrix $\Lambda^{k+1}_{i,\mathscr{G}^{[\eta,k]}}$ is positive definite for each $i \in \mathscr{S}$, we have that $\xi(k) = 0$ if and only if $(G^{[\eta]}(k) - G^{[\eta-1]}(k))X_i^{[\eta-1]}(k) = 0$ for all $i \in \mathscr{S}$. In this case, *Step 3* assures that $G^{[\eta]}(k) = G^{[\eta-1]}(k)$, which in turn implies that $\xi(k) = 0$ if and only if $G^{[\eta]}(k) = G^{[\eta-1]}(k)$.

Finally, the result of Theorem 2.2 then follows by summing up (30) with respect to k, i.e.,

$$J_{\mathscr{G}^{[\eta]}} - J_{\mathscr{G}^{[\eta-1]}} = \sum_{k=0}^{N-1}(J_{\mathscr{G}^{[\eta,k]}} - J_{\mathscr{G}^{[\eta,k+1]}}) = -\sum_{k=0}^{N-1}\xi(k) \le 0,$$

which shows the monotone non-increasing property of the cost sequence $J_{\mathscr{G}^{[\eta]}}$, $\eta = 0, 1, \ldots$. As a byproduct, we have that $J_{\mathscr{G}^{[\eta-1]}} > J_{\mathscr{G}^{[\eta]}}$ whenever $G^{[\eta]}(k) \ne G^{[\eta-1]}(k)$, so that the limit $\lim_{\eta\to\infty} G^{[\eta]}(k)$ exists for every $k = 0, \ldots, N - 1$. This argument completes the proof of Theorem 2.2.   □

2.2.1 Remarks

The results of this monograph can be quite easily extended to the scenario of clustered observation of the Markov state [3], in which one observes the variable ψ_k taking values in the set $\mathscr{S} = \{1, \ldots, \sigma\}$ and satisfying $\psi_k = i$ whenever $\theta_k \in \mathscr{S}_i$, where $\mathscr{S}_i$, $0 \le i \le \sigma$, forms a partition of $\mathscr{S}$. For example, if $\mathscr{S} = \{1, \ldots, 4\}$, $\mathscr{S}_1 = \{1\}$ and $\mathscr{S}_2 = \{2, \ldots, 4\}$, then $\psi_k = 2$ means that $\theta_k \in \{2, \ldots, 4\}$. Note that, if we define the function

$$\psi(i) = \sum_{j=1}^{\sigma} j \, \mathbb{1}_{\{i \in \mathscr{S}_j\}},$$

then $\psi_k = \psi(\theta_k)$ a.s. We assume that the controller is in the form $u_k = G(k, \psi(\theta_k))x_k$, hence the closed loop structure is now given by $A_i(k) := A_i + B_i G(k, \psi(i))$, for all $i \in \mathcal{S}$ and $k = 0, \dots, N - 1$, and the necessary condition for optimality reads as

$$\sum_{i=1}^{\sigma} [(R_i + B_i' \mathscr{E}_i(L(k+1))B_i)G(k, \psi(i)) + B_i' \mathscr{E}_i(L(k+1))A_i]X_i(k) = 0. \tag{31}$$

The algorithm is altered accordingly by substituting (13) by (31).

One interesting feature of (31) is that, assuming the cardinality of $\mathcal{S}_j$ is one for some $0 \le j \le \sigma$, that is, $\mathcal{S}_j = \{r\}$ for some $0 \le r \le \sigma$, then $\psi(\ell) = j$ only when $\ell = r$, allowing to obtain from (31) an analytical expression for $G(k, \psi(r))$

$$G(k, \psi(r)) = -(R_r + B_r' \mathscr{E}_r(L(k+1))B_r)^{-1} B_r' \mathscr{E}_r(L(k+1))A_r,$$

irrespectively of $X(k)$ and the other gains $G(k, \psi(i))$ in (31). In one extreme, the case when the mode is observed (complete observation) can be retrieved by setting $\mathcal{S}_j = \{j\}$, $0 \le j \le \sigma$. In this situation, $\psi(i) = i$ and the optimal gain is given by $G(k, i) = -(R_i + B_i' \mathscr{E}_i(L(k+1))B_i)^{-1} B_i' \mathscr{E}_i(L(k+1))A_i$. The algorithm converges in one iteration, and (16) is now equivalent to the well known Riccati difference equation for the jump linear quadratic problem [1, Chap. 4]. This also serves as an illustration that the dependence of the gains on the second moment matrices $X(k)$ in (31) is not a drawback of the methodology in this monograph, it is a feature of the considered partial observation problem. In the other extreme, (31) and (13) are equivalent if one sets $\mathcal{S}_1 = \mathcal{S}$.

2.3 Numerical Example

This section presents an adapted example from [4], which consists of a continuous-time uncertain system characterized by four different operating points. In [4], a time discretization was performed, leading to four discrete-time linear systems given by

$$A_i = \begin{bmatrix} a_{11}^i & a_{12}^i & a_{13}^i \\ a_{21}^i & a_{22}^i & a_{23}^i \\ 0 & 0 & 0.2231 \end{bmatrix}, \quad B_i = \begin{bmatrix} b_1^i \\ b_2^i \\ 0.7769 \end{bmatrix},$$

$$Q_i = I, \ R_i = 1, \ F_i = 0, \ i = 1, 2, 3, 4,$$

where parameters a_{ij}^i and b_i^i are as listed in [4]. We set $N = 4$, $x_0 = [-0.27\ 1.2\ 2.1]'$, $\mu_0 = [0.25\ 0.25\ 0.25\ 0.25]'$, $H_i = I$, $i = 1, 2, 3, 4$, and $\mathbb{P} = [p_{ii} = 0.88, p_{ij} = 0.04, \forall i, j \in \mathcal{N}, i \neq j]$. We shall assume here that the system can jump from one operating point to another, according to a Markov chain, thus forming a MJLS. We employ two distinct covariance matrix to observe the numerical sensitivity in the example. The following matrices are adopted: *Case 1:* $\Sigma = 0.25I$; *Case 2:* $\Sigma = I$.

Table 1 Optimal feedback gains and minimal cost

	Case 1			Case 2		
K^0	0.024	−0.1067	−0.1867	0.0247	−0.1101	−0.1927
K^1	0.1377	0.3069	−0.3235	0.0858	0.0255	−0.1772
K^2	−0.0626	−0.0158	−0.32	−0.0413	−0.0229	−0.1509
K^3	−0.0805	−0.0875	−0.249	−0.0772	−0.0952	−0.2354
$J_{\mathbf{K}}^{N}$		284.57			373.47	

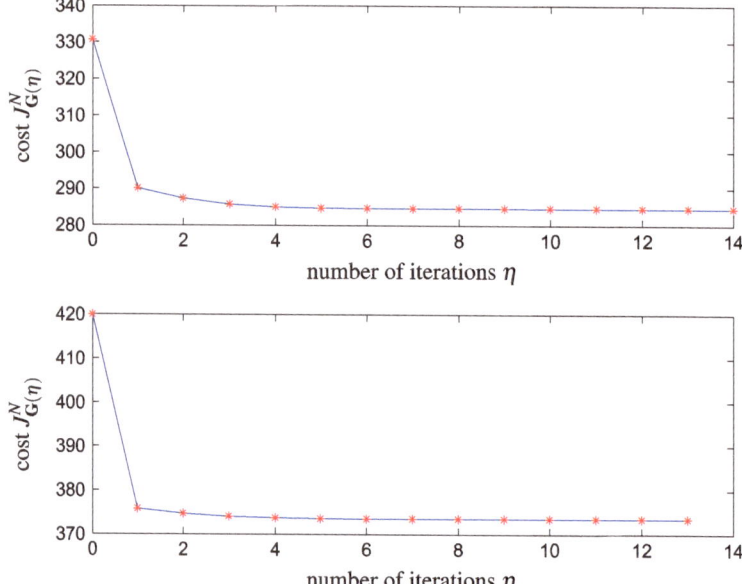

Fig. 2 Cost Evolution for *Case 1* (*above*) and for *Case 2* (*below*)

We use the *MATLAB* software to implement the algorithm proposed in Sect. 2.2. We evaluate, for each case, the algorithm with 10,000 distinct initial sequences $\mathbf{G}(0) \in \mathscr{K}$, and in every simulation the algorithm converges to the same minimal cost. This result confirms the Theorem 2.2, and it is a strong indication that the minimum achieved is the global minimum, since the result is independent of the initial choice of $\mathbf{G}(0)$.

The results obtained for this example, for minimal cost $J_{\mathbf{K}}^{N}$ and optimal sequence $\{K^0, K^1, K^2, K^3\}$, are presented in Table 1.

Figure 2 shows the cost evolution $J_{\mathbf{G}(\eta)}^{N}$ versus number of iterations η, for *Case 1* e *Case 2*. We can see that the value of the minimal cost in *Case 2* is greater than *Case 1*, due to a greater magnitude of the covariance matrix of the former.

Fig. 3 Laboratory DC
Motor testbed used in the
experiments of Sect. 2.4

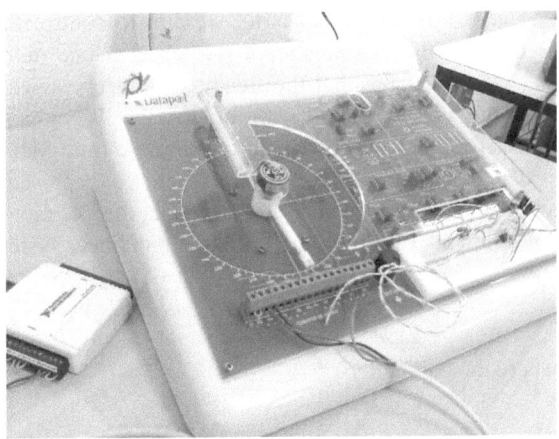

2.4   Experimental Results from a DC Motor Device

This section presents a real implementation of the underlying Markovian controller
for a DC motor device. In practical terms, we use the results of the previous section
to derive a strategy to control the speed of a real DC motor device subject to abrupt
failures. The equipment is altered to take these failures into account according to a
prescribed Markov chain.

The experimental testbed is based on the DC Motor Module 2208, made up by
Datapool Eletronica Ltda, Brazil, using a National Instruments USB-6008 data acqui-
sition card to perform a physical link with the computer, see Fig. 3. The computer
calls MATLAB software to implement physically the controller and it makes use of
the gain sequence that was precomputed offline from Theorem 2.2.

It is known that the dynamics of DC motors can be satisfactorily represented by
second order linear systems [5–7]. In this case, the two system state variables are the
angular velocity of the motor shaft and the electrical current consumed by the motor,
which are represented in this project, respectively, by v_k and i_k, $k \geq 0$. In practice, to
measure the angular velocity, we use the manufacturer-provided tachogenerator that
produces voltage proportional to the speed of the shaft; and to measure the electric
current, we introduce in series with the motor a simple circuit composed by a shunt
resistor connected with a pre-amplifier signal stage. First-order analog filters are
used in the circuit to reduce high-frequency noise from the experimental data. The
experiments of this project are conducted with a sampling period of 15.6118 ms.

Abrupt failures on the power transmitted to the shaft play an important role in the
speed of motors, and this fact motivates us to adjust the apparatus in order to impose
power failures therein. Namely, we force the DC motor device to run under three
distinct operation modes, i.e., the normal, low, and medium power modes, and these
switching modes are programmed to occur according to a homogeneous Markov
chain.

Under this failure scenario, we aim to control the speed of the DC motor so as to track the constant input reference of one radian per second. As a byproduct, we can assure that the steady-state error vanishes to zero. In fact, to accomplish this goal in practice, we modify the PI compensator schematic suggested in [8, Sect. 10.7.3] to cope with the discrete-time MJLS. As a result, by setting the system state as $x_k \equiv [v_k \ i_k \ x_{3,k}]'$ (where $x_{3,k}$ represents the integrative term written as a discrete sum), we are able to model the DC motor device subject to failures as the following discrete-time Markov jump linear system:

$$x_{k+1} = A_{\theta_k} x_k + B_{\theta_k} u_k + H_{\theta_k} w_k + \Gamma_{\theta_k} r_k, \quad k \geq 0, \tag{32}$$

where the parameters

$$A_i = \begin{bmatrix} a_{11}^{(i)} & a_{12}^{(i)} & 0 \\ a_{21}^{(i)} & a_{22}^{(i)} & 0 \\ a_{31}^{(i)} & 0 & a_{33}^{(i)} \end{bmatrix}, \quad B_i = \begin{bmatrix} b_1^{(i)} \\ b_2^{(i)} \\ 0 \end{bmatrix}, \quad \Gamma_i = \begin{bmatrix} 0 \\ 0 \\ \gamma^{(i)} \end{bmatrix}, \quad H_i = h^{(i)}, \quad i = 1, 2, 3.$$

are given in Tables 2 and 3. The sequence $\{w_k\}$ on $\mathscr{R}^2$ represents an i.i.d. noise sequence with zero mean and covariance matrix equal to the identity matrix, $\{r_k\}$ on $\mathscr{R}$ denotes the tracking reference signal, and $\{u_k\}$ on $\mathscr{R}$ stands for the controller.

The design objective of this project is to control the speed of the real DC motor device when sudden power failures occur. The practical experiment in the laboratory testbed implements the controller in the linear state-feedback form

$$u_k = G(k)x_k, \quad \forall k \geq 0. \tag{33}$$

In the control design, we set the model (32) and (33) with $r(k) \equiv 0$ to get a matrix gain sequence $\mathbf{G} = \{G(0), \ldots, G(N-1)\}$ from Theorem 2.2 satisfying the necessary optimal condition of Theorem 2.1. This strategy is purposeful to improve attenuation of the real input disturbances $\omega(\cdot)$ with fast transient response for tracking

Table 2 Parameters of the discrete-time MJLS representing a real DC motor device as in Sect. 2.4

Parameters	$a_{11}^{(i)}$	$a_{12}^{(i)}$	$a_{21}^{(i)}$	$a_{22}^{(i)}$	$a_{31}^{(i)}$	$a_{33}^{(i)}$
$i = 1$	−0.479908	5.1546	−3.81625	14.4723	0.139933	−0.925565
$i = 2$	−1.60261	9.1632	−0.5918697	3.0317	0.0740594	−0.43383
$i = 3$	0.634617	0.917836	−0.50569	2.48116	0.386579	0.0982194

Table 3 Parameters of the discrete-time MJLS representing a real DC motor device as in Sect. 2.4

Parameters	$b_1^{(i)}$	$b_2^{(i)}$	$h^{(i)}$	$\gamma^{(i)}$
$i = 1$	5.87058212	15.50107	0.1	0.11762727
$i = 2$	10.285129	2.2282663	0.1	−0.1328741
$i = 3$	0.7874647	1.5302844	1	0.1632125

Table 4 Parameters for the experimental testbed of Sect. 2.4

Parameters	$q_{11}^{(i)}$	$q_{12}^{(i)}$	$q_{22}^{(i)}$	$q_{33}^{(i)}$	R_i
$i = 1$	0.24	0.61	2.1	0.7	2
$i = 2$	0.8	−0.512	0.676	0.1	2
$i = 3$	0	0	0	0	1000

problems, see [7, 9, 10] for further details regarding deterministic systems. As a consequence, these specifications can be taken into account in our practical experiments designed for the tracking reference $r_k \equiv 1$. Indeed, we will see in the sequence that G engenders an interesting tracking behavior for the speed of the DC Motor device when failures happen.

To perform the experiments, we set $N = 1800$, $\pi_0 = [1\ 0\ 0]'$,

$$
Q_i = \begin{bmatrix} q_{11}^{(i)} & q_{12}^{(i)} & 0 \\ q_{12}^{(i)} & q_{22}^{(i)} & 0 \\ 0 & 0 & q_{33}^{(i)} \end{bmatrix}, \text{ and } F_i = \mathbf{0}_{2\times2}, \quad i = 1, 2, 3,
$$

with values shown in Table 4.

The task of defining precisely the value of the stochastic matrix $\mathbb{P}$ may be cumbersome in some circumstances [11–15]. In this project, however, we are able to define $\mathbb{P}$ precisely as

$$
\mathbb{P} = \begin{bmatrix} 0.84 & 0.07 & 0.09 \\ 0.24 & 0.75 & 0.01 \\ 0.11 & 0.08 & 0.81 \end{bmatrix}.
$$

We can see in Fig. 4 the experimental and simulated data of the angular velocity and electric current for some realization of the Markovian process. Notice in the figure that the experimental and simulated data tend to overlap each other, which is a strong indication that the MJLS model (32)–(33) provides a good representation of the DC motor device subject to power failures. In addition, one can see that the DC Motor speed v_k follows the tracking reference $r_k \equiv 1$ with success, even though the power failures tend to deviate it from its reference target. The figure also presents the states of the Markov chain with respect to the normal ($\theta_k = 1$), low ($\theta_k = 2$), and medium ($\theta_k = 3$) power modes associated with the evolution of the system trajectory.

To clarify the influence of abrupt power failures on the DC motor device in practice, we carry out a Monte Carlo-based experiment. The idea of the Monte Carlo experimentation is to operate the DC Motor device to work out one thousand distinct random experiments, and the corresponding outcome is then used to obtain the mean and standard deviation of both the angular velocity and electric current of the device, see Fig. 5 for a pictorial representation. It is noteworthy that even in the real scenario of failures, the designed controller is able to drive with success the mean value of the DC Motor speed to the tracking reference value of one radian per second. The experimental values of the standard deviation are bounded and this indicates that the stochastic system is stable, cf. [1, Chap. 3], [16].

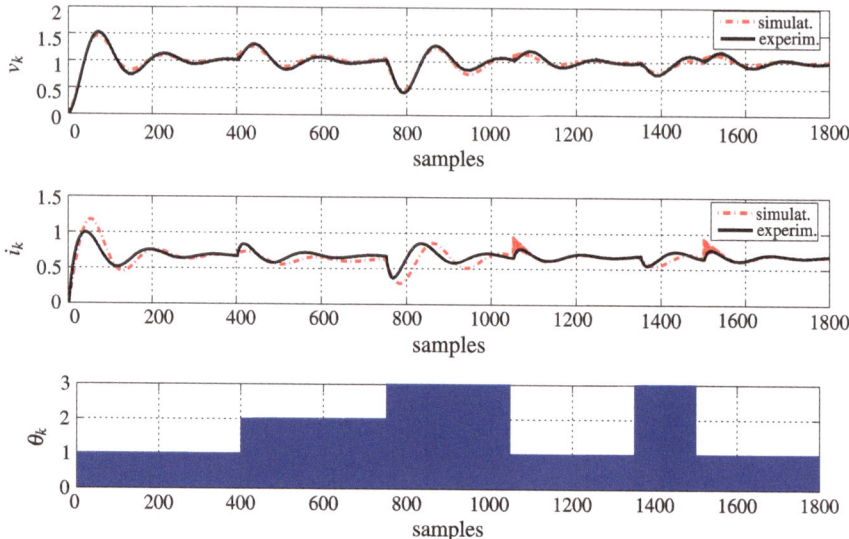

Fig. 4 A sample-path of the angular velocity and electric current obtained from both real and simulated data using the MJLS control strategy of Theorem 2.2. The corresponding state of the Markov chain is depicted in the third picture

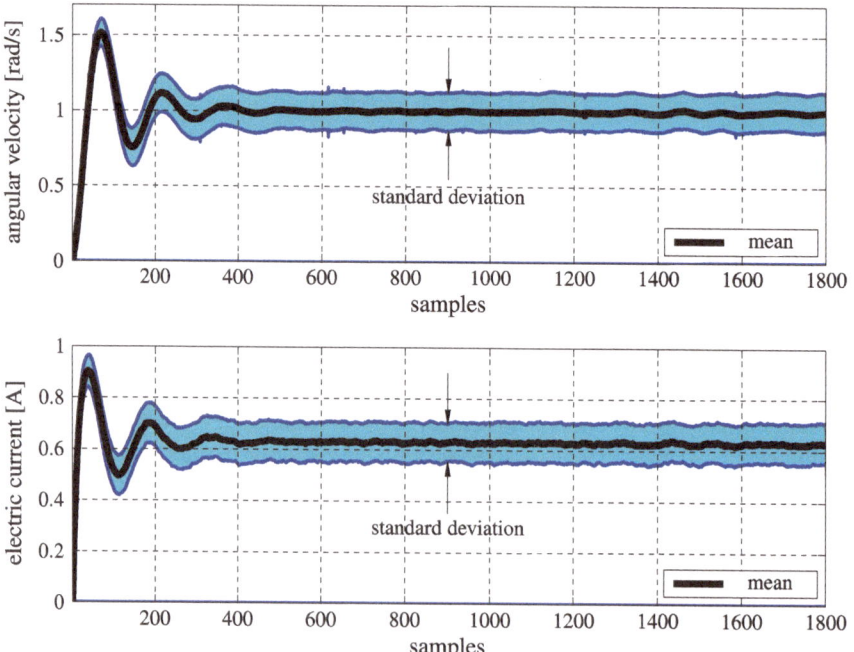

Fig. 5 Mean and standard deviation values of the angular velocity and electric current obtained from Monte Carlo practical experiments. The MJLS control strategy of Theorem 2.2 is used to generate the corresponding experimental data

3 Finite-Time Control Problem: Descendent Methods

For sake of notational simplicity, we assume hereafter no noise in the system (1), that is, (1) equals

$$x_{k+1} = A_{\theta_k} x_k + B_{\theta_k} u_k, \quad \forall k \geq 0, \ x_0 \in \mathscr{R}^r, \ \theta_0 \sim \pi_0. \tag{34}$$

Seeking for simplicity and aiming at practical control applications, we assume that the control law is in the linear static state-feedback format with no mode observation as follows.

$$u_k = G x_k, \quad k \geq 0. \tag{35}$$

Notice that the gain G is stationary, hence it does not depend on the time index k.

The optimization control problem we deal with is that of finding some matrix G that minimizes (1) subject to (1) and (35). Formally, if we let $J_N(G)$ be the cost (1) for a given G, then we recast the optimization control problem as follows.

$$G^* = \arg \min_G J_N(G). \tag{36}$$

To the best of the authors' knowledge, there is no method to compute the optimal solution for the control problem in (36). A drawback for finding the optimal solution of (36) is the fact that the nonlinear functional $J_N(G)$ may be non-convex (see Sect. 2.4). A tentative method to overcome this difficulty, aiming at the optimal solution, is to employ optimization techniques borrowed from the literature, although these techniques are able to guarantee stationary points only (i.e., local minimum or saddle points).

The main contribution of this section is twofold. First, we derive the expression of the gradient of the optimization problem in (36). Second, we recall some optimization techniques from the literature to compare their efficiency on achieving the solution of (36) for a particular control problem.

3.1 Preliminaries

Define the inner product on the space $\mathbb{M}^{r,s}$ as

$$\langle U, V \rangle = \sum_{i=1}^{\sigma} \text{tr}\{U_i' V_i\}, \quad \forall V, U \in \mathbb{M}^{r,s}, \tag{37}$$

and the Frobenius norm $\|U\|_2^2 = \langle U, U \rangle$.

If $f : \mathscr{M}^{s,r} \mapsto \mathscr{R}$ is a differentiable function on the domain $\mathscr{M}^{s,r}$, we denote its partial derivative by $\partial f(G)/\partial G$ whenever $G \in \mathscr{M}^{s,r}$. Let $\text{tr}\{\cdot\}$ denote the trace operator. We now recall some derivative rules for the trace operator. Considering

U, V, Z, and G as matrices with compatible dimensions, we have [17, Sect. 10.3.2]

$$\frac{\partial \operatorname{tr}\{UGV\}}{\partial G} = U'V', \quad \frac{\partial \operatorname{tr}\{UG'V\}}{\partial G} = VU, \quad \frac{\partial \operatorname{tr}\{UGVG'Z\}}{\partial G} = U'Z'GV' + ZUGV. \quad (38)$$

Let us define the conditional second moment matrix of the system state x_k, $k \geq 0$, as

$$X_i(k) = \mathrm{E}[x_k x_k' \, \mathbb{1}_{\{\theta_k=i\}}], \quad \forall i \in \mathscr{S}, \ \forall k \geq 0,$$

where $\mathbb{1}_{\{\cdot\}}$ stands for the Dirac measure. Using this definition, we can write the identity [1, p. 31]

$$\mathrm{E}_{x_0, \pi_0}[x_k'(Q_{\theta_k} + G'R_{\theta_k}G)x_k]$$

$$= \sum_{i=1}^{\sigma} \operatorname{tr}\left\{(Q_i + G'R_iG)\mathrm{E}_{x_0, \pi_0}[x_k x_k' \, \mathbb{1}_{\{\theta_k=i\}}]\right\} = \langle Q + G'RG, X(k)\rangle, \quad \forall k \geq 0.$$

Thus, the Nth horizon cost function $J_N(x_0, \pi_0)$ as in (1) can be written equivalently as

$$J_N(G) = \sum_{k=0}^{N} \langle Q + G'RG, X(k)\rangle. \quad (39)$$

To evaluate precisely the cost $J_N(G)$ as in (39), let us set $X(k) = \{X_1(k), \ldots, X_\sigma(k)\} \in \mathbb{S}^{n0}$, $k \geq 0$, and notice that it satisfies the recurrence [1, Proposition 3.1]

$$X(k+1) = \mathscr{D}\big((A + BG)X(k)(A + BG)'\big), \quad \forall k \geq 0, \quad (40)$$

with $X_i(0) = \pi_i(0)x_0 x_0'$ for each $i \in \mathscr{S}$.

Finally, to complete the definition of recurrences required in the next results, let us consider the sets $W(k) \in \mathbb{S}^{r0}$, $k = 0, \ldots, N$, generated as follows.

$$W(k+1) = (A + BG)'\mathscr{E}(W(k))(A + BG), \quad k = 0, \ldots, N-1, \ \text{and}$$

$$W(0) = Q + G'RG. \quad (41)$$

3.2 Main Results

Lemma 3.1 *For each $k = 0, \ldots, N$, there holds*

$$\frac{\partial \langle Q + G'RG, X(k)\rangle}{\partial G}$$

$$= 2\left(\sum_{j=0}^{\sigma} R_j GX_j(k) + \sum_{\ell=0}^{k-1}\sum_{i=0}^{\sigma} B_i'\mathscr{E}_i\big(W(k-1-\ell)\big)(A_i + B_iG)X_i(\ell)\right).$$

Proof To prove the main result, it is necessary to introduce some auxiliary results. To begin with, notice from the formulas (38) that we can write

$$U, V \in \mathscr{S}^r \Rightarrow \frac{\partial \operatorname{tr}\{U(A_i + B_i G)V(A_i + B_i G)'\}}{\partial G} = 2B_i'U(A_i + B_i G)V,$$

$$i = 1, \ldots, \sigma. \tag{42}$$

Let us now turn our attention to the recurrence (40). If we set $k = 1$ in (40), one can verify that

$$X_{i_1}(1) = \sum_{i_0=1}^{\sigma} p_{i_0 i_1}(A_{i_0} + B_{i_0} G)X_{i_0}(0)(A_{i_0} + B_{i_0} G)', \quad i_1 = 1, \ldots, \sigma.$$

With $k = 2$ in (40), we have

$$X_{i_2}(2) = \sum_{i_1=1}^{\sigma} \sum_{i_0=1}^{\sigma} p_{i_0 i_1} p_{i_1 i_2}(A_{i_1} + B_{i_1} G)(A_{i_0} + B_{i_0} G)$$
$$\times X_{i_0}(0)(A_{i_0} + B_{i_0} G)'(A_{i_1} + B_{i_1} G)', \quad i_2 = 1, \ldots, \sigma.$$

Proceeding similarly with $k = \ell + 1$ in (40), we obtain

$$X_{i_{\ell+1}}(\ell + 1) = \sum_{i_\ell=1}^{\sigma} \cdots \sum_{i_0=1}^{\sigma} \left(p_{i_0 i_1} \cdots p_{i_\ell i_{\ell+1}}(A_{i_\ell} + B_{i_\ell} G) \cdots (A_{i_0} + B_{i_0} G) \right.$$
$$\left. \times X_{i_0}(0)(A_{i_0} + B_{i_0} G)' \cdots (A_{i_\ell} + B_{i_\ell} G)' \right), \quad i_{\ell+1} = 1, \ldots, \sigma.$$

$$\tag{43}$$

Combining (37) and (43), we obtain the identity

$$\frac{\partial \langle Q + G'RG, X(\ell + 1) \rangle}{\partial G} = \sum_{i_{\ell+1}=1}^{\sigma} \frac{\partial}{\partial G} \operatorname{tr}\left\{(Q_{i_{\ell+1}} + G'R_{i_{\ell+1}} G)X_{i_{\ell+1}}(\ell + 1)\right\}$$

$$= \sum_{i_{\ell+1}=1}^{\sigma} \cdots \sum_{i_0=1}^{\sigma} p_{i_0 i_1} \cdots p_{i_\ell i_{\ell+1}} \left[\frac{\partial}{\partial G} \operatorname{tr}\left\{(Q_{i_{\ell+1}} + G'R_{i_{\ell+1}} G) \right.\right.$$
$$\times (A_{i_\ell} + B_{i_\ell} G) \cdots (A_{i_0} + B_{i_0} G)X_{i_0}(0)$$
$$\left.\left. \times (A_{i_0} + B_{i_0} G)' \cdots (A_{i_\ell} + B_{i_\ell} G)' \right\} \right]. \tag{44}$$

On the other hand, the derivative chain rule [17, Sect. 10.3.1] states that

$$
\frac{\partial \langle Q + G'RG, X(\ell+1) \rangle}{\partial G} = \frac{\partial \langle \overbrace{Q + G'RG}^{\text{variable}}, \overbrace{X(\ell+1)}^{\text{fixed}} \rangle}{\partial G} + \frac{\partial \langle \overbrace{Q + G'RG}^{\text{fixed}}, \overbrace{X(\ell+1)}^{\text{variable}} \rangle}{\partial G}. \tag{45}
$$

The first expression in the right-hand side of the equality (45) is identical to (see (38))

$$
\sum_{i=1}^{\sigma} \frac{\partial \operatorname{tr}\{\overbrace{Q_i + G'R_iG}^{\text{variable}}, \overbrace{X_i(\ell+1)}^{\text{fixed}}\}}{\partial G} = \sum_{i=1}^{\sigma} 2R_iGX_i(\ell+1). \tag{46}
$$

To evaluate the second term in the right-hand side of (45), we start with (44) taking $Q_{i_{\ell+1}} + G'R_{i_{\ell+1}}G$ as a fixed term. The derivative chain rule will be useful in this calculation. Indeed, the idea behind the derivative chain rule is to consider (44) with $(A_{i_0} + B_{i_0}G)$ as variable and all of the other terms fixed, and after this we take $(A_{i_1} + B_{i_1}G)$ as variable and all of the other terms fixed, and so on until the evaluation of the term $(A_{i_\ell} + B_{i_\ell}G)$ is accomplished.

Let us now start this procedure. Assume that $U \in \mathscr{S}^r$ and $V \in \mathscr{S}^r$ are fixed and defined in (42) as

$$
U = (A_{i_1} + B_{i_1}G)' \cdots (A_{i_\ell} + B_{i_\ell}G)'(Q_{i_{\ell+1}} + G'R_{i_{\ell+1}}G)(A_{i_\ell} + B_{i_\ell}G) \cdots (A_{i_1} + B_{i_1}G),
$$
$$
V = X_{i_0}(0).
$$

Thus, the term inside the brackets of (44) equals

$$
\frac{\partial}{\partial G} \operatorname{tr}\{U(A_{i_0} + B_{i_0}G)V(A_{i_0} + B_{i_0}G)'\},
$$

which yields

$$
2B_{i_0}' \Bigg[(A_{i_1} + B_{i_1}G)' \cdots (A_{i_\ell} + B_{i_\ell}G)'(Q_{i_{\ell+1}} + G'R_{i_{\ell+1}}G)
$$
$$
\times (A_{i_\ell} + B_{i_\ell}G) \cdots (A_{i_1} + B_{i_1}G) \Bigg](A_{i_0} + B_{i_0}G)X_{i_0}(0).
$$

Substituting this expression into (44), we obtain

$$
\sum_{i_0=1}^{\sigma} 2B_{i_0}' \Bigg[\sum_{i_1=1}^{\sigma} p_{i_0 i_1}(A_{i_1} + B_{i_1}G)' \cdots \sum_{i_\ell=1}^{\sigma} p_{i_{\ell-1}i_\ell}(A_{i_\ell} + B_{i_\ell}G)'
$$
$$
\times \sum_{i_\ell=1}^{\sigma} p_{i_\ell i_{\ell+1}}(Q_{i_{\ell+1}} + G'R_{i_{\ell+1}}G)(A_{i_\ell} + B_{i_\ell}G) \cdots (A_{i_1} + B_{i_1}G) \Bigg](A_{i_0} + B_{i_0}G)X_{i_0}(0).
$$

Notice that the term inside the brackets is identical to $\mathscr{E}_{i_0}(W(\ell))$. Hence, when $(A_{i_0} + B_{i_0}G)$ is variable and all of the other terms remain fixed, we get that

$$\frac{\partial \langle Q + G'RG, X(\ell + 1) \rangle}{\partial G} = \sum_{i_0=1}^{\sigma} 2B'_{i_0} \mathscr{E}_{i_0}(W(\ell))(A_{i_0} + B_{i_0}G)X_{i_0}(0).$$

Let us now assume that the term $(A_{i_1} + B_{i_1}G)$ is variable and all of the others are fixed. Since (44) can be rewritten as

$$\sum_{i_{\ell+1}=1}^{\sigma} \cdots \sum_{i_1=1}^{\sigma} p_{i_1 i_2} \cdots p_{i_\ell i_{\ell+1}} \left[\frac{\partial}{\partial G} \text{tr} \left\{ (Q_{i_{\ell+1}} + G'R_{i_{\ell+1}}G) \right. \right.$$
$$\left. \left. \times (A_{i_\ell} + B_{i_\ell}G) \cdots (A_{i_1} + B_{i_1}G)X_{i_1}(1)(A_{i_1} + B_{i_1}G)' \cdots (A_{i_\ell} + B_{i_\ell}G)' \right\} \right],$$

one can repeat the previous reasoning, taking $(A_{i_1} + B_{i_1}G)$ as variable and all of the other terms fixed, to show that

$$\frac{\partial \langle Q + G'RG, X(\ell + 1) \rangle}{\partial G} = \sum_{i_1=1}^{\sigma} 2B'_{i_1} \mathscr{E}_{i_1}(W(\ell - 1))(A_{i_1} + B_{i_1}G)X_{i_1}(1).$$

Finally, summing up the elements from this argument, we obtain

$$\frac{\partial \langle Q + G'RG, \overbrace{X(\ell + 1)}^{\text{variable}} \rangle}{\partial G} = \sum_{i_0=1}^{\sigma} 2B'_{i_0} \mathscr{E}_{i_0}(W(\ell))(A_{i_0} + B_{i_0}G)X_{i_0}(0)$$
$$+ \sum_{i_1=1}^{\sigma} 2B'_{i_1} \mathscr{E}_{i_1}(W(\ell - 1))(A_{i_1} + B_{i_1}G)X_{i_1}(1)$$
$$\vdots$$
$$+ \sum_{i_\ell=1}^{\sigma} 2B'_{i_\ell} \mathscr{E}_{i_\ell}(W(0))(A_{i_\ell} + B_{i_\ell}G)X_{i_\ell}(\ell). \quad (47)$$

(the underbrace for "fixed" spans $\langle Q + G'RG$.)

The desired result then follows from (45), (46), and (47). □

The next result is an immediate consequence of Lemma 3.1 and the expression for the cost in (39).

Theorem 3.1 Let $\varphi : \mathscr{M}^{s,r} \mapsto \mathscr{M}^{s,r}$ be the gradient of the cost $J_N(G)$ as in (39). Then, it satisfies

$$\frac{\partial J_N(G)}{\partial G} = \varphi(G), \quad (48)$$

where

$$\varphi(G) := 2 \sum_{k=0}^{N} \left(\sum_{j=0}^{\sigma} R_j G X_j(k) + \sum_{\ell=0}^{k-1} \sum_{i=0}^{\sigma} B_i' \mathscr{E}_i \big(W(k-1-\ell) \big) (A_i + B_i G) X_i(\ell) \right), \quad (49)$$

and $X(k) \in \mathbb{S}^{r_0}$ *and* $W(k) \in \mathbb{S}^{r_0}$ *satisfy* (40) *and* (41), *respectively.*

The next result is immediate from Theorem 3.1 and [18, Corollary p. 185].

Corollary 3.1 (Necessary optimal condition) *If* $\bar{G} \in \mathscr{M}^{s,r}$ *is a local minimum, then* $\varphi(\bar{G}) = 0$.

3.3 Methodology

The aim of this section is to describe the methodology we use to evaluate the necessary optimal condition of Corollary 3.1. For this purpose, let us consider the gradient of (39), evaluated at a point **G**, as

$$\varphi(\mathbf{G}) = \left. \frac{\partial J_N(G)}{\partial G} \right|_{G=\mathbf{G}}. \quad (50)$$

We focus our study on conjugate gradient and quasi-Newton methods [18–24], and all of these algorithms are based on the following three steps.

Step 1. Choose $\varepsilon > 0$ and some initial point $\mathbf{G}_0$. Set $k = 0$.

Step 2. Find an appropriate descent direction $\mathbf{d}_k$ and compute the scalar α_k such that

$$\alpha_k := \arg\min_{\alpha > 0} J_N(\mathbf{G}_k + \alpha \mathbf{d}_k).$$

Step 3. Set $\mathbf{G}_{k+1} = \mathbf{G}_k + \alpha_k \mathbf{d}_k$ and $k = k + 1$. Return to Step 2 if $\|\varphi(\mathbf{G}_k)\| \geq \varepsilon$.

Notice that *Steps 1–3* produce a sequence of points $\mathbf{G}_0, \mathbf{G}_1, \ldots, \mathbf{G}_k, \ldots$, and hopefully we can choose a subsequence $\mathbf{G}_{n_0}, \mathbf{G}_{n_1}, \ldots, \mathbf{G}_{n_k}, \ldots$ from it such that

$$\varphi(\mathbf{G}_{n_k}) \to 0 \quad \text{as} \quad k \to \infty. \quad (51)$$

An accumulation point $\mathbf{G}_\infty := \lim_{k \to \infty} \mathbf{G}_{n_k}$ satisfies the necessary optimal condition for (39) (Corollary 3.1), i.e.,

$$\varphi(\mathbf{G}_\infty) = 0.$$

As a consequence, $\mathbf{G}_\infty$ realizes a local minimum or a saddle point for (39). Notice that a local minimum may coincide with the global one, and in this case we have $\mathbf{G}_\infty = G^*$.

We select in our analysis the following ten optimization algorithms due to their wide use in practice, good speed of convergence, and general acceptance in the literature:

- Steepest descent (SD), see [19, Sect. 8.5], [18, Sect. 8.6];
- Davidon-Fletcher-Powell (DFP), see [19, Sect. 8.6], [24, Sect. 5.1];
- Fletcher-Reeves (FR), see [19, Sect. 8.6], [18, p. 278];
- Zangwill (Z), see [19, Sect. 8.6];
- Broyden-Fletcher-Goldfarb-Shanno (BFGS), see [24, Sect. 5.4.1];
- Hestenes-Stiefel (HS), see [24, Sect. 4.2.1];
- Perry (P), see [22, 23];
- Dai-Yuan (DY), see [25];
- Liu-Storey (LS), see [26].

Remark 3.1 The expression of the gradient function $\varphi(\cdot)$ as in (49) is the key to evaluate the conjugate gradient and quasi-Newton methods (SD), (DFP), (FR), (Z), (BFGS), (HR), (P), (DY), and (LS). The sequence of descent directions

$$(\mathbf{d}_0, \mathbf{d}_1, \ldots, \mathbf{d}_k, \ldots)$$

in *Step 2* requires the computation of the gradient $\varphi(\mathbf{G}_k)$ for every point $\mathbf{G}_k \in \mathscr{M}^{s,r}$, $k \geq 0$, cf. [18–20, 24].

3.4 Numerical Evaluations

The main goal of this section is to illustrate the efficiency of the ten selected optimization algorithms (SD), (DFP), (FR), (Z), (BFGS), (HR), (P), (DY), and (LS).

In the numerical evaluations, we consider the same values used in the example in Sect. 2.4. In addition, we use the expressions in (39) and (40) to evaluate the optimization algorithms (SD), (DFP), (FR), (Z), (BFGS), (HR), (P), (DY), and (LS) according to the *Steps 1–3* with initial point $\mathbf{G}_0 = [0\ 0\ 0]$. All of these algorithms converge successfully to the same point $\mathbf{G}_\infty$ given by

$$\mathbf{G}_\infty = [0.0104\ 0.10832\ -0.28469]. \tag{52}$$

One can check that $\varphi(\mathbf{G}_\infty) \simeq 0$, so that $\mathbf{G}_\infty$ is a candidate for a local minimum according to Corollary 3.1.

To evaluate the efficiency of the optimization algorithms, we check the number of iterations required by each algorithm to converge to the stationary point $\mathbf{G}_\infty$ within a tolerance of $\varepsilon = 10^{-3}$ (i.e., $\|\varphi(\mathbf{G}_\infty)\| < \varepsilon$).

Despite the fact that the number of iterations required for the convergence vary drastically from one method to another, a relevant conclusion we can take is that all of the algorithms converges successfully to the same point of minimum (Table 5). In

Table 5 Results obtained from an evaluation of nine selected optimization algorithms according to the numerical example of Sect. 3.4

Method	Num. Iter.	$\|\varphi(\mathbf{G}_k)\|$	$J_N(\mathbf{G}_k)$
(SD)	786	9.79277×10^{-4}	$2.7198609489860 \times 10^{-2}$
(DFP)	45	8.79546×10^{-4}	$2.7198609489800 \times 10^{2}$
(FR)	106	6.09576×10^{-4}	$2.7198609489785 \times 10^{2}$
(Z)	620	9.21772×10^{-4}	$2.7198609489860 \times 10^{2}$
(BFGS)	99	8.80098×10^{-4}	$2.7198609489745 \times 10^{2}$
(HS)	141	9.34734×10^{-4}	$2.7198609489762 \times 10^{2}$
(P)	101	9.58184×10^{-4}	$2.7198609489763 \times 10^{2}$
(DY)	173	4.77101×10^{-4}	$2.719860948974 \times 10^{2}$
(LS)	294	9.93084×10^{-4}	$2.719860948988 \times 10^{2}$

The results indicate that the DFP algorithm is the quickest in the convergence to a local minimum

addition, the (DFP) algorithm is the quickest one to reach a local minimum point, while (SD) is the slowest one. As a byproduct, $\mathbf{G}_\infty$ is a stabilizing gain in the mean square sense [1, Theorem 3.9, p. 36].

3.5 Concluding Remarks

In this chapter, we have shown two methods to calculate the optimal solution of the Markov jump control problem. Using controllers with no mode observation, we have developed two strategies to calculate the necessary optimality conditions (i.e., point of local minimizers): variational method and gradient descendent method.

Both methods guarantee local minimizers for the control problem, and they will be useful in the design of a method to calculate the long-run average cost.

In Sect. 2.4, the derived control strategy satisfying the optimal condition is applied in practice to control the speed of a real DC motor device subject to abrupt power failures. The contribution of this approach is reinforced by the Monte Carlo experiment, which shows that even in the case with sudden power failures, the proposed MJLS controller with no mode observation is able to control the speed of the DC motor device.

4 Next Chapter: Approximation Method

Recall the Markov jump linear system (see (1))

$$x_{k+1} = A_{\theta_k} x_k + B_{\theta_k} u_k + H_{\theta_k} w_k, \quad \forall k \geq 0, \ x_0 \in \mathscr{R}^r, \ \theta_0 \sim \pi_0, \qquad (53)$$

where x_k, u_k, and w_k represent the state, control input, and noisy disturbances, respectively.

The control applies in the linear state feedback with no Markov mode observation, i.e., it assumes the format

$$u(k) = g(k)x(k), \quad \forall k \geq 0, \tag{54}$$

where $g(k)$ stands for a matrix of dimension $m \times n$. Substituting (54) into (53) yields

$$x_{k+1} = (A_{\theta_k} + B_{\theta_k}g(k))x(k) + H_{\theta_k}w(k), \quad \forall k \geq 0. \tag{55}$$

Given any sequence of gains $\mathbf{g} = \{g(0), g(1), \ldots\}$, we can calculate the long-run average cost associated with the system (53), as follows:

$$J(\mathbf{g}) = \limsup_{N \to \infty} \frac{1}{N} \sum_{k=0}^{N-1} E[x(k)'Q_{\theta_k}x(k) + u(k)'R_{\theta_k}u(k)] \quad \text{s.t.} \quad (54). \tag{56}$$

Let $\mathbf{G}$ be the set made up of all admissible sequences $\mathbf{g} = \{g(0), g(1), \ldots\}$. The control problem we are interested in solving is defined next.

$$J^* = \min_{\mathbf{g} \in \mathbf{G}} J(\mathbf{g}) \quad \text{s.t.} \quad (53) \text{ and } (54).$$

The next chapters advance in a method to compute the optimal value J^*. The main idea behind the method is as follows. Consider the Nth stage control problem

$$J_N^* = \min_{\mathbf{g} \in \mathbf{G}} \left(\sum_{k=0}^{N-1} E\left[x(k)'(Q_{\theta_k} + g(k)'R_{\theta_k}g(k))x(k) \right] \right). \tag{57}$$

With the method to be developed in the next chapter, we present conditions to assure the validity of the approximation

$$J_N^*/N \to J^* \quad \text{when} \quad N \to \infty, \tag{58}$$

for any initial condition x_0 and $\pi(0)$.

Note that if any algorithm attains the global minimizer $\mathbf{g}_N^* = \{g^*(0), \ldots, g^*(N-1)\}$ of J_N^* from (57), then $\mathbf{g}_N^*$ can be used to calculate J^* through (58).

To the best of the authors' knowledge, there is no algorithm that assuredly computes the global minimizer $\mathbf{g}_N^*$; however, the algorithm of *Steps 1–4* generates a *candidate* for the global minimizer $\mathbf{g}_N^*$. In other words, this algorithm generates a local minimum, which could differ from the global minimum. To assure that the local and global minimum coincide, we introduce the next conjecture.

Conjecture 4.1 The optimal control problem in (57) has a unique minimum, and the corresponding local minimizers coincide with the global minimizer.

Theorem 4.1 *Under Conjecture 4.1, a gain sequence $\mathbf{g}_N^* = \{g^*(0), \ldots, g^*(N-1)\}$ realizes the minimum in (57) if and only if $\mathbf{g}_N^*$ satisfies necessary optimality conditions for J_N^*.*

Remark 4.1 The result of Theorem 4.1 guarantees that any gain satisfying necessary optimality conditions is the optimal solution of the control problem in (57). This result is important because it allows us to use the algorithm of *Steps 1–4* to solve the Nth stage problem in (57), see Theorem 2.2 in connection.

References

1. O.L.V. Costa, M.D. Fragoso, R.P. Marques, *Discrete-Time Markovian Jump Linear Systems* (Springer, New York, 2005)
2. J.B.R. do Val, T. Başar, Receding horizon control of jump linear systems and a macroeconomic policy problem. J. Econ. Dyn. Control **23**, 1099–1131 (1999)
3. J.B.R. do Val, J.C. Geromel, A.P. Gonçalves, The H_2 control for jump linear systems cluster observations of the Markov state. Automatica **38**, 343–349 (2002)
4. J.C. Geromel, P.L.D. Peres, S.R. Souza, $\mathscr{H}_2$-guaranteed cost control for uncertain discrete-time linear systems. Intern. J. Control **57**, 853–864 (1993)
5. W. Leonhard, *Control of Electrical Drives*, 3rd edn. (Springer, New York, 2001)
6. A. Rubaai, R. Kotaru, Online identification and control of a DC motor using learning adaptation of neural networks. IEEE Trans. Ind. Appl. **36**(3), 935–942 (2000)
7. M. Ruderman, J. Krettek, F. Hoffmann, T. Bertram, Optimal state space control of DC motor, in *Proceedings of the 17th IFAC World Congress* (Seoul, Korea, 2008), pp. 5796–5801
8. C.L. Phillips, R.D. Harbor, *Feedback Control Systems*, 3rd edn. (Prentice Hall, Upper Saddle River, 1996)
9. E. Assunção, C.Q. Andrea, M.C.M. Teixeira, H_2 and H_∞-optimal control for the tracking problem with zero variation. IET Control Theory Appl. **1**(3), 682–688 (2007)
10. A. Kojima, S. Ishijima, LQ preview synthesis: optimal control and worst case analysis. IEEE Trans. Automat. Control **44**(2), 352–357 (1999)
11. X. Luan, F. Liu, P. Shi, Finite-time stabilization of stochastic systems with partially known transition probabilities. J. Dyn. Syst. Meas. Control Trans. ASME. **133**(1), 14504–14510 (2011)
12. R.C.L.F. Oliveira, A.N. Vargas, J..B.R. do Val, P.L.D. Peres, Robust stability, H$_2$ analysis and stabilisation of discrete-time Markov jump linear systems with uncertain probability matrix. Int. J. Control, **82**(3), 470 – 481 (2009)
13. Y. Yin, P. Shi, F. Liu, Gain scheduled PI tracking control on stochastic nonlinear systems with partially known transition probabilities. J. Frankl. Inst. **348**, 685–702 (2011)
14. Y. Yin, P. Shi, F. Liu, J.S. Pan, Gain-scheduled fault detection on stochastic nonlinear systems with partially known transition jump rates. Nonlinear Anal. Real World Appl. **13**, 359–369 (2012)
15. L. Zhang, E. Boukas, L. Baron, H.R. Karimi, Fault detection for discrete-time Markov jump linear systems with partially known transition probabilities. Int. J. Control **83**(8), 1564–1572 (2010)
16. A.N. Vargas, J.B.R. do Val, Average cost and stability of time-varying linear systems. IEEE Trans. Autom. Control, **55**, 714–720 (2010)
17. H. Lütkepohl, *Handbook of Matrices* (Wiley, Hoboken, 1996)

18. D.G. Luenberger, Y. Ye. *Linear and Nonlinear Programming*, 3rd edn. (Springer, Heidelberg, 2010)
19. M.S. Bazaraa, H.D. Sherali, C.M. Shetty, *Nonlinear Programming: Theory and Algorithms*, 3rd edn. (Wiley-Interscience, Hoboken, 2006)
20. D.P. Bertsekas. *Nonlinear Programming* (Athena Scientific, Nashua, 1999)
21. Y.-H. Dai, Convergence properties of the BFGS algoritm. SIAM J. Optim. **13**(3), 693–701 (2002)
22. I.E. Livieris, P. Pintelas, Globally convergent modified Perry's conjugate gradient method. Appl. Math. Comput. **218**(18), 9197–9207 (2012)
23. A. Perry, A modified conjugate gradient algorithm. Oper. Res. **26**(6), 1073–1078 (1978)
24. W. Sun, Y.X. Yuan, *Optimization Theory and Methods: Nonlinear Programming* (Springer, Heidelberg, 2006)
25. Y.H. Dai, Y. Yuan, A nonlinear conjugate gradient method with a strong global convergence property. SIAM J. Optim. **10**(1), 177–182 (1999)
26. Y. Liu, C. Storey, Efficient generalized conjugate gradient algorithms, part 1: Theory. J. Optim. Theory Appl. **69**, 129–137 (1991)

Approximation of the Optimal Long-Run Average-Cost Control Problem

1 Preliminaries

Consider a discrete-time stochastic linear system defined in a filtered probability space $(\Omega, \mathscr{F}, \{\mathscr{F}_k\}, P)$ as follows.

$$x_{k+1} = A(g_k)x_k + Ew_k, \quad g_k \in \mathscr{G}, \quad x_0 \in \mathbb{R}^n, \forall k = 0, 1, \ldots, \qquad (1)$$

where x_k and $w_k, k = 0, 1, \ldots$ are processes taking values, respectively, in $\mathbb{R}^n$ and $\mathbb{R}^q$, which represent the system state, and additive noisy input, respectively. The noisy input $\{w_k\}$ forms an iid process with zero mean and covariance matrix equal to the identity for each $k \geq 0$. The matrix E, of dimension $n \times q$, is given. The variable g_k, at the kth stage, represents the *control action* and belongs to a prescribed set $\mathscr{G}$. We assume that A is a continuous operator, possibly nonlinear, that maps $\mathscr{G}$ to the space of real matrices of dimension $n \times n$.

Let us consider the cost of N stages

$$J_N = \sum_{k=0}^{N-1} \mathrm{E}[x_k' Q(g_k) x_k], \qquad (2)$$

where $\mathrm{E}[\cdot]$ denotes the mathematical expectation and Q is a given operator that maps $\mathscr{G}$ to the space of nonnegative symmetric matrices of dimension $n \times n$.

Associated with (1), we consider the second moment of the system state x_k as

$$X_k = \mathrm{E}[x_k x_k'], \quad \forall k \geq 0. \qquad (3)$$

The control action $g_k \in \mathscr{G}$ applied in (1) and (2) is assumed to be a function of the second moment only, i.e., it takes the *deterministic feedback* form $g_k = f_k(X_k)$

© The Author(s) 2016

A.N. Vargas et al., *Advances in the Control of Markov Jump Linear Systems with No Mode Observation*, SpringerBriefs in Control, Automation and Robotics, DOI 10.1007/978-3-319-39835-8_3

for each $k \geq 0$. Note that this special form suggests simplicity of solutions, since it turns valid the identity

$$E[x_k' Q(g_k) x_k] = \langle Q(g_k), X_k \rangle,$$

where $\langle \cdot, \cdot \rangle$ represents the usual Frobenius inner product. As a matter of fact, the main reason for adopting this particular feedback structure is that the system state x_k may not be available for g_k. This situation occurs, for instance, if g_k is a *gain matrix*. Moreover, if g_k is taken to be a gain matrix in (1)–(2), then some important control problems can be represented by means of (1)–(2) indeed involving feedback. One interesting situation that can be handled in that way is the simultaneous state-feedback control problem (see [1–6] for a small account). Recall that a simultaneous state-feedback control system with different operating points can be represented as

$$\varphi_i(k+1) = (A_i + B_i g(k))\varphi_i(k) + E_i \omega_i(k), \quad i = 1, \ldots, \sigma, \tag{4}$$

where $g(k)$ is a design gain matrix that does not depend on the mode i, and $\varphi_i(\cdot)$ and $\omega_i(\cdot)$ represent the simultaneous system state and additive noise input for the ith mode, respectively. The cost of N stages is given by

$$J_N = \sum_{k=0}^{N-1} \sum_{i=1}^{\sigma} E[\varphi_i(k)'(Q_i + g(k)' R_i g(k))\varphi_i(k))], \tag{5}$$

where the positive-semidefinite matrices Q_i and R_i are given.

We state that (4)–(5) can be rewritten as (1)–(2). Indeed, if we set

$$\dim(A_i) = n \times n, \quad \dim(B_i) = n \times r, \quad \text{and} \quad \dim(E_i) = n \times q,$$

then the admissible set $\mathscr{G}$ is $\mathbb{R}^{r \times n}$ and

$$A(g) = \text{diag}(A_1 + B_1 g, \ldots, A_\sigma + B_\sigma g),$$
$$Q(g) = \text{diag}(Q_1 + g' R_1 g, \ldots, Q_\sigma + g' R_\sigma g), \quad \forall g \in \mathscr{G},$$

and $E = \text{diag}(E_1, \ldots, E_\sigma)$. The claimed correspondence between (4)–(5) and (1)–(2) follows by simply stacking the simultaneous system state and additive noise input, respectively, in the form

$$x_k = \begin{bmatrix} \varphi_1(k) \\ \vdots \\ \varphi_\sigma(k) \end{bmatrix} \in \mathbb{R}^{\sigma n}, \quad \text{and} \quad w_k = \begin{bmatrix} \omega_1(k) \\ \vdots \\ \omega_\sigma(k) \end{bmatrix} \in \mathbb{R}^{\sigma q}.$$

Hence, we can conclude that the study of the model in (1)–(2) may provide insights on how to solve some relevant control problems, see Sect. 3 for an application in the average cost simultaneous control problem.

The approximating control problem we deal with is as follows. The feedback functions f_k, $k \geq 0$, specify a policy $\mathbf{f} = \{f_0, \ldots, f_k, \ldots\}$ (see [7, 8]). If $\mathbf{F}$ denotes the set of all feasible policies $\mathbf{f}$, then the associated problem of N stages is defined as

$$J_N^* = \min_{\mathbf{f} \in \mathbf{F}} J_N.$$

The long-run average cost is defined as

$$\overline{J} = \limsup_{N \to \infty} \frac{J_N}{N}.$$

and the corresponding average cost problem is

$$\overline{J}^* = \min_{\mathbf{f} \in \mathbf{F}} \overline{J}.$$

The main contribution of this chapter is on determining conditions under which

$$J_N^*/N \to \overline{J}^* \quad \text{as} \quad N \to \infty. \tag{6}$$

We recall that results similar to (6) are available in the Markov decision process (MDP) literature, for instance, under an equicontinuous assumption on $\{J_N^*\}$ and Borel state space in [8, Chap. 5, p. 102]; a countable action space together with some technical assumptions on the relative value functions in [9, Theorem 3.1]; and a bounded cost-by-stage condition and countable state space in [10, Corollary 4.2]. On the other hand, when compared with MDP, our approach is simpler to verify because we take advantage of the particular structure of (1)–(2). To assure (6), we require a controllability condition together with a condition that the last element of the second moment trajectory, corresponding to the optimal N-stage cost, does not increase faster than N, see Theorem 3.1 in connection.

The system (1)–(2) associated with the long-run average cost was studied in [11–13]. These papers, basically, present conditions to assure the existence of an optimal stationary policy for the average cost problem. Here, we advance on the investigation by deriving an approximation method to evaluate the optimal average cost by means of finite N stage optimal costs.

The chapter is organized as follows. Section 2 presents the necessary notation, definitions, assumptions, and the main result. The main result concerning the approximation method is presented in Theorem 3.1. Section 3 is dedicated to apply the approximation result to the simultaneous state-feedback control problem. Section 4 contains the proofs, and some concluding remarks are presented in Sect. 5.

2 Notation and Main Results

The real and natural numbers are denoted by $\mathbb{R}$ and $\mathbb{N}$, respectively. The set of nonnegative real numbers is denoted by $\mathbb{R}_+$, and $\mathbb{R}^{n,m}$ is used to represent the space of all $n \times m$ real matrices. The superscript $'$ indicates the transpose of a matrix. Let $\mathbb{S}_+^n$ be the closed convex cone $\{U \in \mathbb{R}^{n,n} : U = U' \geq 0\}$; and $\| \cdot \|$ will denote either the standard Euclidean norm in $\mathbb{R}^n$ or the Frobenius norm for matrices. We say that a matrix sequence $\{U_k; k \geq 0\}$ is bounded if $\sup_{k \in \mathbb{N}} \|U_k\| < \infty$.

The following definitions and conventions will apply throughout this chapter.

(i) $\mathscr{X}$ and $\mathscr{G}$ are given sets referred to as *state space* and *control space*, respectively. In particular, we assume that $\mathscr{X} \subseteq \mathbb{S}_+^n$ and $\mathscr{G}$ are Borel spaces.

(ii) For each $X \in \mathscr{X}$, there is given a nonempty measurable subset $\mathscr{G}(X)$ of $\mathscr{G}$. The set $\mathscr{G}(X)$ represents the set of *feasible controls* or *actions* when the system is in state $X \in \mathscr{X}$, and with the property that the graph

$$\text{Gr} := \{(X, g) | X \in \mathscr{X}, g \in \mathscr{G}(X)\} \tag{7}$$

of feasible state-action pairs is measurable.

(iii) (inf-compactness [8, p. 28]). Let $Q : \mathscr{G} \to \mathbb{S}_+^n$ be a lower semi-continuous function. The one-stage cost functional $\mathscr{C} : \text{Gr} \to \mathbb{R}_+$ is defined as follows:

$$\mathscr{C}(X, g) = \langle X, Q(g) \rangle, \quad \forall (X, g) \in \text{Gr}. \tag{8}$$

Moreover, for each $X \in \mathscr{X}$ and $\lambda \in \mathbb{R}_+$, the set $\{g \in \mathscr{G}(X) | \mathscr{C}(X, g) \leq \lambda\}$ is compact.

(iv) A *policy* $\mathbf{f} = \{f_0, f_1, \ldots\}$ is made up by a sequence of measurable functions $f_k : \mathscr{X} \to \mathscr{G}, k \geq 0$, and the set of all policies is denoted by $\mathbf{F}$. Elements of $\mathbf{F}$ of the form $\mathbf{f} = \{f, f, \ldots\}$ are referred to as *stationary policies*.

From the assumption on the process $\{w_k\}$, $k \geq 0$, and for a given policy $\mathbf{f} = \{f_k\} \in \mathbf{F}$, the second moment matrix $X_k \in \mathscr{X}$ from (3) satisfies the recurrence (cf. [14, Chap. 2])

$$X_{k+1} = A(g_k) X_k A(g_k)' + \Sigma, \quad \forall k \geq 0, \quad \forall X_0 = X \in \mathscr{X}, \tag{9}$$

with $\Sigma := EE'$, where the control obeys the rule

$$g_k = f_k(X_k), \quad \forall k \geq 0. \tag{10}$$

Sometimes we use the notation $X_k^{(\mathbf{f})}$ to stress that the recurrence (9) depends on a specific $\mathbf{f}$. Accordingly, we represent the kth stage cost by

$$\mathscr{C}_k^{(\mathbf{f})} := \mathscr{C}(X_k^{(\mathbf{f})}, g_k) = \langle X_k^{(\mathbf{f})}, Q(g_k) \rangle, \quad \forall k \geq 0.$$

The cost of N stages is defined by

$$J_N(\mathbf{f}, X) := \sum_{k=0}^{N-1} \mathscr{C}_k^{(\mathbf{f})}, \quad \forall N \geq 1, \tag{11}$$

and the corresponding Nth stage control problem is of finding a sequence of feedback control functions $\psi_N^* := \{f_0, \ldots, f_{N-1}\}$ such that

$$J_N^*(X) := J_N(\psi_N^*, X) = \inf_{\mathbf{f} \in \mathbf{F}} J_N(\mathbf{f}, X). \tag{12}$$

The existence of ψ_N^*, $N = 1, 2, \ldots$ is assured by the inf-compactness assumption, see [8, Chap. 3].

The long-run average cost is defined by

$$\overline{J}(\mathbf{f}, X) := \limsup_{N \to \infty} \frac{1}{N} \sum_{k=0}^{N-1} \mathscr{C}_k^{(\mathbf{f})}, \tag{13}$$

and the corresponding control problem is of finding a policy $\mathbf{f}^* \in \mathbf{F}$ such that

$$\overline{J}^*(X) := \overline{J}(\mathbf{f}^*, X) = \inf_{\mathbf{f} \in \mathbf{F}} \overline{J}(\mathbf{f}, X). \tag{14}$$

A policy $\mathbf{f}^*$ satisfying (14) is referred to as average cost optimal.

2.1 Discounted Criterion and the Main Result

One of our assumption is based on the behavior of the discounted criteria with discounts tending to one. Formally, for each $\alpha \in (0, 1)$, the discounted criterion we shall deal with is defined as

$$V_\alpha(\mathbf{f}, X) := \sum_{k=0}^{\infty} \alpha^k \mathscr{C}_k^{(\mathbf{f})}, \quad \forall \mathbf{f} \in \mathbf{F}, \quad \forall X_0 = X \in \mathscr{X}, \tag{15}$$

where α denotes the *discount factor*. The associated control problem is of finding $\mathbf{f}_\alpha^* \in \mathbf{F}$ such that

$$V_\alpha^*(X) := V_\alpha(\mathbf{f}_\alpha^*, X) = \inf_{\mathbf{f} \in \mathbf{F}} V_\alpha(\mathbf{f}, X), \quad \forall X_0 = X \in \mathscr{X}. \tag{16}$$

The existence of a policy $\mathbf{f}_\alpha^* \in \mathbf{F}$ that satisfies (16) is assured by the inf-compactness assumption [8, Chap. 3].

The next definition simplifies the notation to be used in the sequel.

Definition 3.1 For some policy $\mathbf{f} = \{f_k\} \in \mathbf{F}$ and $X_0 = \Sigma$ fixed, let $\{X_k\}$ and $\{g_k\}$ be defined as in (9) and (10), respectively. We define the transition matrix from Σ, $\Phi^{(\mathbf{f})} : \mathbb{N} \to \mathbb{M}^{n,n}$ as

$$\Phi^{(\mathbf{f})}(k) = A(g_{k-1}) \ldots A(g_0), \quad k \geq 1,$$

with $\Phi^{(\mathbf{f})}(0)$ being the identity matrix. Similarly, we define $Q_k^{(\mathbf{f})} = Q(g_k)$ for each $k \geq 0$.

Let us now consider the following assumption.

Assumption 3.1 (*Controllability to the origin*, [12]). The following two statements hold.

(i) There exists a natural number N and a finite sequence of control actions $\{g_{c_0}, \ldots, g_{c_N}\}$ such that
$$A(g_{c_N}) \ldots A(g_{c_0}) = 0.$$

(ii) There exist a constant $M \geq 0$ such that

$$\limsup_{\alpha \uparrow 1} \sum_{k=0}^{\infty} \alpha^k \left\langle \Phi^{(\mathbf{f}_\alpha^*)}(k)' Q_k^{(\mathbf{f}_\alpha^*)} \Phi^{(\mathbf{f}_\alpha^*)}(k) , S_c \right\rangle \leq M,$$

where $\mathbf{f}_\alpha^* \in \mathbf{F}$ satisfies (16) and the matrix S_c is defined as

$$S_c = A(g_{c_0}) \Sigma A(g_{c_0})' + A(g_{c_1}) A(g_{c_0}) \Sigma A(g_{c_0})' A(g_{c_1})'$$
$$+ \ldots + A(g_{c_{N-1}}) \ldots A(g_{c_0}) \Sigma A(g_{c_0})' \ldots A(g_{c_{N-1}})'. \quad (17)$$

Now, we are able to present the main result of this chapter.

Theorem 3.1 *Suppose that Assumption 3.1 holds. Let ψ_N^*, $N \geq 1$, be a sequence of feedback functions satisfying (12). If*

$$\lim_{N \to \infty} \|X_N^{(\psi_N^*)}\|/N = 0, \quad (18)$$

then there exists a constant ρ (which does not depend on $X \in \mathcal{X}$) such that

$$J_N^*(X)/N \to \rho = \bar{J}^*(X) \text{ as } N \to \infty, \quad (19)$$

for all $X \in \mathcal{X}$.

The proof of Theorem 3.1 will be given in Sect. 4.

3 Numerical Example

For sake of a numerical evaluation, we recast the uncertain system presented in [15] as a simultaneous state feedback one in (4). We consider the simultaneous system in (4) with four different operating points

$$A_i = \begin{bmatrix} a_{11}^i & a_{12}^i & a_{13}^i \\ a_{21}^i & a_{22}^i & a_{23}^i \\ 0 & 0 & 0.2231 \end{bmatrix}, \quad B_i = \begin{bmatrix} b_1^i \\ b_2^i \\ 0.7769 \end{bmatrix}, \quad i = 1, 2, 3, 4,$$

where the parameters a_{ij}^i and b_i^i are as listed in [15]. We adopt $E_i = [0\,0\,1]'$, $Q_i = I$, $R_i = 1$, and $x_i(0) = [-0.27\,1.2\,2.1]'$ for each $i = 1, 2, 3, 4$.

Now, we show that Assumption 3.1 holds. Indeed, after rewriting the simultaneous system in the form of (1)–(2), we obtain

$$A(g)E = \mathrm{diag}(A_1 + B_1g, \ldots, A_4 + B_4g) \cdot \mathrm{diag}(E_1, \ldots, E_4).$$

If

$$g_c = [0\,0\,-0.2231/0.7769],$$

then $A(g_c)E = 0$, so that

$$A(g_c)EE'A(g_c)' = A(g_c)\Sigma A(g_c)' = 0.$$

It follows from (17) that $S_c = 0$ and Assumption 3.1 is satisfied trivially.

To evaluate the result of Theorem 3.1, it remains to show that the limit in (18) holds true. For this purpose, we use a variational method based on the one described in [16, 17] to compute a matrix gain sequence that is candidate for the optimal solution of the Nth stage control problem $J_N^*(X_0)$. In other words, the variational method guarantees local minimizers only but they may coincide with the global ones.

Hereafter, we use the conjecture that the local minimizer is also a global one in order to compute $J_N^*(X_0)$ and the corresponding optimal gain sequence. From Fig. 1, we can see that the sequence $\{\|X_N^{(\psi_N^*)}\|\}$ is bounded, thus the limit in (18) holds. Thus, under the conjecture, Theorem 3.1 assures that the finite horizon optimal cost $J_N^*(X_0)$ asymptotically approximates the optimal long-run average cost $\rho = \bar{J}^*(X)$ for any $X \in \mathscr{X}$, i.e., for each $\varepsilon > 0$ there holds

$$|J_N^*(X_0)/N - \bar{J}^*(X)| < \varepsilon, \quad \forall X \in \mathscr{X},$$

for sufficiently large values of N. Moreover, the optimal value from this approximation is (see Fig. 2 in connection)

$$\rho = \bar{J}^*(X) = 202.05, \quad \forall X \in \mathscr{X}.$$

Fig. 1 Plot of the sequence $\{\|X_N^{(\psi_N^*)}\|\}$ in the example of Sect. 3

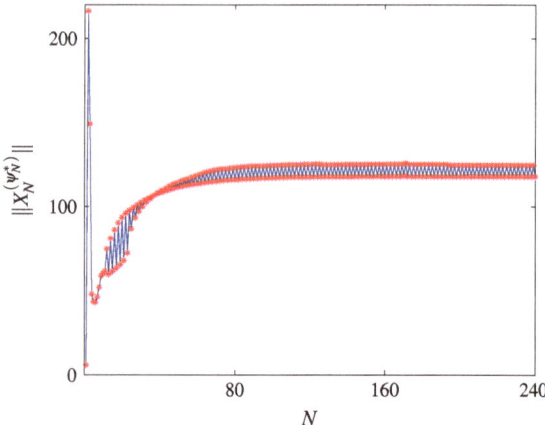

Fig. 2 Plot of the optimal cost $J_N^*(X_0)$ divided by the number of stages N in the example of Sect. 3.

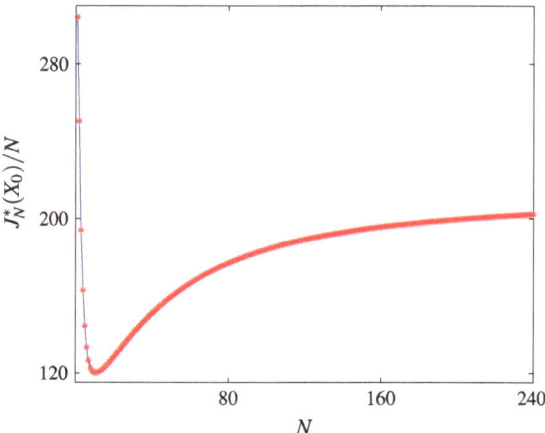

4 Proof of Theorem 3.1

To prove Theorem 3.1, we introduce some preliminary results.

Proposition 3.1 ([12]) *If Assumption 3.1 holds, then there exists a sequence of discount factors $\alpha_n \uparrow 1$ such that:*

(H_1) There exists a constant $c > 0$ such that

$$(1 - \alpha_n)V_{\alpha_n}^*(0) \leq c, \quad \forall n \geq 0. \tag{20}$$

(H_2) There exist two positive constants c_1, c_2 such that

$$0 \leq h_{\alpha_n}(X) \leq c_1\|X\| + c_2, \quad \forall X \in \mathscr{X}, \quad \forall n \geq 0, \tag{21}$$

where $h_{\alpha_n}(X) := V^*_{\alpha_n}(X) - V^*_{\alpha_n}(0)$.

We note that (H_1) and (H_2) are conditions frequently required in Markov decision processes to deal with average cost problems, see for instance [8, Chap. 5], [9, 18, 19].

Proposition 3.2 ([8, Theorem 5.4.3, p. 88, and Remark 4, p. 95]). *If both (H_1) and (H_2) are valid, then*

 (i) *There exist a sequence of discount factors $\alpha_n \uparrow 1$ and a constant ρ such that*

$$\lim_{n\to\infty} (1 - \alpha_n)V^*_{\alpha_n}(X) = \rho, \quad \forall X \in \mathcal{X}. \tag{22}$$

(ii) *By defining the function $h : \mathcal{X} \to \mathbb{R}_+$ as*

$$h(X) := \liminf_{n\to\infty} h_{\alpha_n}(X), \quad \forall X \in \mathcal{X}, \tag{23}$$

we have $0 \leq h(X) \leq c_1 \|X\| + c_2$, for all $X \in \mathcal{X}$.
(iii) *There exists a feedback function $f \in \mathcal{F}$ such that*

$$\begin{aligned}
\rho + h(X) &\geq \min_{g \in \mathcal{G}(X)} \left[\mathcal{C}(X, g) + h\big(A(g)XA(g)' + \Sigma\big) \right] \\
&= \mathcal{C}(X, f(X)) \\
&\quad + h\big(A(f(X))XA(f(X))' + \Sigma\big), \quad \forall X \in \mathcal{X}.
\end{aligned} \tag{24}$$

Moreover, the stationary policy $\mathbf{f} = \{f, f, \ldots\} \in \mathbf{F}$ satisfies

$$\rho = \overline{J}^*(X) = \overline{J}(\mathbf{f}, X), \quad \forall X \in \mathcal{X}.$$

The next well-known result will be useful in the sequel.

Proposition 3.3 ([8, Theorem 4.2.3, p. 46], [19, Theorem 2.1]) *Under inf-compactness and stabilizability, there holds*

$$V^*_\alpha(X) = \min_{g \in \mathcal{G}(X)} \left[\mathcal{C}(X, g) + \alpha V^*_\alpha\big(A(g)XA(g)' + \Sigma\big) \right],$$

for each $\alpha \in (0, 1)$ and $X \in \mathcal{X}$.

The next important result states that Assumption 3.1 assures an average cost optimality equation.

Lemma 3.1 *If Assumption 3.1 holds, then*

$$\rho + h(X) = \min_{g \in \mathcal{G}(X)} \left[\mathcal{C}(X, g) + h\big(A(g)XA(g)' + \Sigma\big) \right], \quad \forall X \in \mathcal{X}, \tag{25}$$

where $\rho > 0$ and $h(\cdot)$ are as in Proposition 3.2.

Proof Let $\alpha_n \uparrow 1$ be the sequence of discount factors satisfying Proposition 3.2. Combining $h_{\alpha_n}(X) = V_{\alpha_n}^*(X) - V_{\alpha_n}^*(0)$ and Proposition 3.3, we can write

$$(1 - \alpha_n)V_{\alpha_n}^*(0) + h_{\alpha_n}(X) = \min_{g \in \mathcal{G}(X)} \left[\mathscr{C}(X, g) + \alpha_n \cdot h_{\alpha_n}\left(A(g)XA(g)' + \Sigma\right)\right],$$

which in turn implies that

$$(1 - \alpha_n)V_{\alpha_n}^*(0) + h_{\alpha_n}(X)$$
$$\leq \mathscr{C}(X, g) + \alpha_n \cdot h_{\alpha_n}\left(A(g)XA(g)' + \Sigma\right), \quad \forall g \in \mathcal{G}. \tag{26}$$

Passing the limit inferior in (26) with respect to $\alpha_n \uparrow 1$, and using (22) and (23), we obtain

$$\rho + h(X) \leq \mathscr{C}(X, g) + h\left(A(g)XA(g)' + \Sigma\right), \quad \forall g \in \mathcal{G}.$$

This implies that

$$\rho + h(X) \leq \min_{g \in \mathcal{G}(X)} \left[\mathscr{C}(X, g) + h\left(A(g)XA(g)' + \Sigma\right)\right],$$

which combined with (24) yield the desired result.

At this point, we are able to introduce the main argument to prove Theorem 3.1.

4.1 Proof of Theorem 3.1 Continued

We now show the result in (19). Applying an induction argument on (25) we see that

$$n\rho + h(X_0) \leq \sum_{k=0}^{n-1} \mathscr{C}_k^{(\mathbf{f})} + h\left(X_n^{(\mathbf{f})}\right), \quad \forall \mathbf{f} \in \mathbf{F}.$$

We have in particular from this inequality that

$$n\rho + h(X_0) \leq J_n^*(X_0) + h\left(X_n^{(\psi_n^*)}\right), \tag{27}$$

where ψ_n^* represents the n-stage optimal policy that satisfies $J_n(\psi_n^*, X_0) = J_n^*(X_0)$. From (H_2) in Proposition 3.1, we have

$$0 \leq h(X_n^{(\psi_n^*)}) \leq c_1 \|X_n^{(\psi_n^*)}\| + c_2. \tag{28}$$

where $h_{\alpha_n}(X) := V^*_{\alpha_n}(X) - V^*_{\alpha_n}(0)$.

We note that (H_1) and (H_2) are conditions frequently required in Markov decision processes to deal with average cost problems, see for instance [8, Chap. 5], [9, 18, 19].

Proposition 3.2 ([8, Theorem 5.4.3, p. 88, and Remark 4, p. 95]). *If both (H_1) and (H_2) are valid, then*

(i) *There exist a sequence of discount factors $\alpha_n \uparrow 1$ and a constant ρ such that*

$$\lim_{n \to \infty} (1 - \alpha_n) V^*_{\alpha_n}(X) = \rho, \quad \forall X \in \mathscr{X}. \tag{22}$$

(ii) *By defining the function $h : \mathscr{X} \to \mathbb{R}_+$ as*

$$h(X) := \liminf_{n \to \infty} h_{\alpha_n}(X), \quad \forall X \in \mathscr{X}, \tag{23}$$

we have $0 \le h(X) \le c_1 \|X\| + c_2$, for all $X \in \mathscr{X}$.

(iii) *There exists a feedback function $f \in \mathscr{F}$ such that*

$$\begin{aligned}
\rho + h(X) &\ge \min_{g \in \mathscr{G}(X)} \left[\mathscr{C}(X, g) + h\big(A(g)XA(g)' + \Sigma\big) \right] \\
&= \mathscr{C}(X, f(X)) \\
&\quad + h\big(A(f(X))XA(f(X))' + \Sigma\big), \quad \forall X \in \mathscr{X}.
\end{aligned} \tag{24}$$

Moreover, the stationary policy $\mathbf{f} = \{f, f, \ldots\} \in \mathbf{F}$ satisfies

$$\rho = \overline{J}^*(X) = \overline{J}(\mathbf{f}, X), \quad \forall X \in \mathscr{X}.$$

The next well-known result will be useful in the sequel.

Proposition 3.3 ([8, Theorem 4.2.3, p. 46], [19, Theorem 2.1]) *Under inf-compactness and stabilizability, there holds*

$$V^*_\alpha(X) = \min_{g \in \mathscr{G}(X)} \left[\mathscr{C}(X, g) + \alpha V^*_\alpha\big(A(g)XA(g)' + \Sigma\big) \right],$$

for each $\alpha \in (0, 1)$ and $X \in \mathscr{X}$.

The next important result states that Assumption 3.1 assures an average cost optimality equation.

Lemma 3.1 *If Assumption 3.1 holds, then*

$$\rho + h(X) = \min_{g \in \mathscr{G}(X)} \left[\mathscr{C}(X, g) + h\big(A(g)XA(g)' + \Sigma\big) \right], \quad \forall X \in \mathscr{X}, \tag{25}$$

where $\rho > 0$ and $h(\cdot)$ are as in Proposition 3.2.

Proof Let $\alpha_n \uparrow 1$ be the sequence of discount factors satisfying Proposition 3.2. Combining $h_{\alpha_n}(X) = V^*_{\alpha_n}(X) - V^*_{\alpha_n}(0)$ and Proposition 3.3, we can write

$$(1 - \alpha_n)V^*_{\alpha_n}(0) + h_{\alpha_n}(X) = \min_{g \in \mathcal{G}(X)} \left[\mathcal{C}(X, g) + \alpha_n \cdot h_{\alpha_n}\left(A(g)XA(g)' + \Sigma\right)\right],$$

which in turn implies that

$$(1 - \alpha_n)V^*_{\alpha_n}(0) + h_{\alpha_n}(X)$$
$$\leq \mathcal{C}(X, g) + \alpha_n \cdot h_{\alpha_n}\left(A(g)XA(g)' + \Sigma\right), \quad \forall g \in \mathcal{G}. \tag{26}$$

Passing the limit inferior in (26) with respect to $\alpha_n \uparrow 1$, and using (22) and (23), we obtain

$$\rho + h(X) \leq \mathcal{C}(X, g) + h\left(A(g)XA(g)' + \Sigma\right), \quad \forall g \in \mathcal{G}.$$

This implies that

$$\rho + h(X) \leq \min_{g \in \mathcal{G}(X)} \left[\mathcal{C}(X, g) + h\left(A(g)XA(g)' + \Sigma\right)\right],$$

which combined with (24) yield the desired result.

At this point, we are able to introduce the main argument to prove Theorem 3.1.

4.1 Proof of Theorem 3.1 Continued

We now show the result in (19). Applying an induction argument on (25) we see that

$$n\rho + h(X_0) \leq \sum_{k=0}^{n-1} \mathcal{C}_k^{(\mathbf{f})} + h\left(X_n^{(\mathbf{f})}\right), \quad \forall \mathbf{f} \in \mathbf{F}.$$

We have in particular from this inequality that

$$n\rho + h(X_0) \leq J_n^*(X_0) + h\left(X_n^{(\psi_n^*)}\right), \tag{27}$$

where ψ_n^* represents the n-stage optimal policy that satisfies $J_n(\psi_n^*, X_0) = J_n^*(X_0)$. From (H_2) in Proposition 3.1, we have

$$0 \leq h(X_n^{(\psi_n^*)}) \leq c_1 \|X_n^{(\psi_n^*)}\| + c_2. \tag{28}$$

The limit in (18) assures that, for each $\varepsilon > 0$, there exists a natural number $n_0(\varepsilon)$ such that

$$n \geq n_0(\varepsilon) \implies (c_1 \|X_n^{(\psi_n^*)}\| + c_2)/n < \varepsilon. \tag{29}$$

Hence, we get from (27)–(29) that

$$n \geq n_0(\varepsilon) \implies \rho < J_n^*(X_0)/n + \varepsilon. \tag{30}$$

On the other hand, it follows from Proposition 3.2 (iii) that there exists a stationary policy $\mathbf{f} = \{f, f, \ldots\}$ such that

$$n\rho + h(X_0) \geq \sum_{k=0}^{n-1} \mathscr{C}_k^{(\mathbf{f})} + h(X_n^{(\mathbf{f})}), \quad \forall n \geq 1.$$

But then, since $\sum_{k=0}^{n-1} \mathscr{C}_k^{(\mathbf{f})} \geq J_n^*(X_0)$ and $h(X_n^{(\mathbf{f})}) \geq 0$, we have

$$n\rho + h(X_0) \geq J_n^*(X_0), \quad \forall n \geq 1.$$

From this inequality, we have that there is a natural number $n_1(\varepsilon)$ such that

$$n \geq n_1(\varepsilon) \implies \rho + \varepsilon > J_n^*(X_0)/n. \tag{31}$$

As a result of combining (30) and (31) we get that

$$n \geq \max\{n_0(\varepsilon), n_1(\varepsilon)\} \implies -\varepsilon < \rho - J_n^*(X_0)/n < \varepsilon,$$

which yields the result in (19). □

5 Concluding Remarks

This chapter presents conditions for which the N-stage optimal cost J_N^*, divided by the number of stages, asymptotically approximates the optimal long-run average cost $\bar{J}^*$ (see Theorem 3.1), i.e.,

$$J_N^*/N \to \bar{J}^* \quad \text{as } N \to \infty.$$

We have indicated that some interesting control problems can be solved in the setup developed here. The possible solution of the problem of simultaneous feedback was shown to satisfy the required assumptions, thus illustrating the usefulness of our approach.

References

1. Y.-Y. Cho, J. Lam, A computational method for simultaneous LQ optimal control design via piecewise constant output feedback. IEEE Trans. Syst. Man Cybern. Part B **31**, 836–842 (2001)
2. G.D. Howitt, R. Luus, Control of a collection of linear systems by linear state feedback control. Int. J. Control **58**(1), 79–96 (1993)
3. R.A. Luke, P. Dorato, C.T. Abdallah, Linear-quadratic simultaneous performance design, in *Proc. American Control Conference* (New Mexico, 1997), pp. 3602–3605
4. J. Lavaei, A.G. Aghdam, Simultaneous LQ control of a set of LTI systems using constrained generalized sampled-data hold functions. Automatica **43**(2), 274–280 (2007)
5. F. Saadatjooa, V. Derhami, S.M. Karbassi, Simultaneous control of linear systems by state feedback. Comput. Math. Appl. **58**(1), 154–160 (2009)
6. J.-L. Wu, T.-T. Lee, Optimal static output feedback simultaneous regional pole placement. IEEE Trans. Syst. Cybern. Part B **35**, 881–893 (2005)
7. D.P. Bertsekas, S.E. Shreve, *Stochastic Optimal Control: The Discrete Time Case* (Athena Scientific, Belmont, 1996)
8. O. Hernández-Lerma, J.B. Lasserre, *Discrete-Time Markov Control Processes: Basic Optimality Criteria* (Springer, New York, 1996)
9. S.P. Meyn, The policy iteration algorithm for average reward Markov decision processes with general state space. IEEE Trans. Autom. Control **42**(12), 1663–1680 (1997)
10. L.I. Sennott, The convergence of value iteration in average cost Markov decision chains. Oper. Res. Lett. **19**, 11–16 (1996)
11. A.N. Vargas, J.B.R. do Val, Minimum second moment state for the existence of average optimal stationary policies in linear stochastic systems, in *Proceeding of American Control Conference* (Baltimore, 2010), pp. 373–377
12. A.N. Vargas, J.B.R. do Val, A controllability condition for the existence of average optimal stationary policies of linear stochastic systems, in *Proceedings of European Control Conference* (Budapest, 2009), pp. 32–37
13. A.N. Vargas, J.B.R. do Val, Average optimal stationary policies: convexity and convergence conditions in linear stochastic control systems, in *Proceedings of 48th IEEE Conference Decision Control and 28th Chinese Control Conference* (Shangai, 2009), pp. 3388–3393
14. B.D.O. Anderson, J.B. Moore, *Optimal Filtering* (Prentice-Hall, Englewood Cliffs, 1979)
15. J.C. Geromel, P.L.D. Peres, S.R. Souza, H_2-guaranteed cost control for uncertain discrete-time linear systems. Int. J. Control **57**, 853–864 (1993)
16. J.B.R. do Val, T. Başar, Receding horizon control of jump linear systems and a macroeconomic policy problem. J. Econ. Dyn. Control **23**, 1099–1131 (1999)
17. A.N. Vargas, J.B.R. do Val, E.F. Costa, Receding horizon control of Markov jump linear systems subject to noise and unobservable state chain, in *Proceedings of 43th IEEE Conference Decision Control* (2004), pp. 4381–4386
18. M. Schal, Average optimality in dynamic programming with general state space. Math. Oper. Res. **18**, 163–172 (1993)
19. A. Arapostathis, V.S. Borkar, E. Fernández-Gaucherand, M.K. Ghosh, S.I. Marcus, Discrete-time controlled Markov processes with average cost criterion: A survey. SIAM J. Control Optim. **31**(2), 282–344 (1993)

Series Editor's Biographies

Tamer Başar is with the University of Illinois at Urbana-Champaign, where he holds the academic positions of Swanlund Endowed Chair, Center for Advanced Study Professor of Electrical and Computer Engineering, Research Professor at the Coordinated Science Laboratory, and Research Professor at the Information Trust Institute. He received the B.S.E.E. degree from Robert College, Istanbul, and the M.S., M.Phil, and Ph.D. degrees from Yale University. He has published extensively in systems, control, communications, and dynamic games, and has current research interests that address fundamental issues in these areas along with applications such as formation in adversarial environments, network security, resilience in cyber-physical systems, and pricing in networks.

In addition to his editorial involvement with these Briefs, Basar is also the Editor in Chief of Automatica, Editor of two Birkhäuser Series on Systems & Control and Static & Dynamic Game Theory, the Managing Editor of the Annals of the International Society of Dynamic Games (ISDG), and member of editorial and advisory boards of several international journals in control, wireless networks, and applied mathematics. He has received several awards and recognitions over the years, among which are the Medal of Science of Turkey (1993); Bode Lecture Prize (2004) of IEEE CSS; Quazza Medal (2005) of IFAC; Bellman Control Heritage Award (2006) of AACC; and Isaacs Award (2010) of ISDG. He is a member of the US National Academy of Engineering, Fellow of IEEE and IFAC, Council Member of IFAC (2011–2014), a past president of CSS, the founding president of ISDG, and president of AACC (2010–2011).

Antonio Bicchi is Professor of Automatic Control and Robotics at the University of Pisa. He graduated from the University of Bologna in 1988 and was a postdoc scholar at M.I.T. A.I. Lab between 1988 and 1990. Following are his main research interests:

- dynamics, kinematics, and control of complex mechanical systems, including robots, autonomous vehicles, and automotive systems;

© The Author(s) 2016
A.N. Vargas et al., *Advances in the Control of Markov Jump Linear Systems with No Mode Observation*, SpringerBriefs in Control, Automation and Robotics, DOI 10.1007/978-3-319-39835-8

- haptics and dextrous manipulation; and theory and control of nonlinear systems, in particular hybrid (logic/dynamic, symbol/signal) systems;
- theory and control of nonlinear systems, in particular hybrid (logic/dynamic, symbol/signal) systems.

He has published more than 300 papers in international journals, books, and refereed conferences.

Professor Bicchi currently serves as the Director of the Interdepartmental Research Center "E. Piaggio" of the University of Pisa, and President of the Italian Association or Researchers in Automatic Control. He has served as Editor in Chief of the Conference Editorial Board for the IEEE Robotics and Automation Society (RAS), and as Vice President of IEEE RAS, Distinguished Lecturer, and Editor for several scientific journals including the *International Journal of Robotics Research, the IEEE Transactions on Robotics and Automation, and IEEE RAS Magazine*. He has organized and co-chaired the first World Haptics Conference (2005), and Hybrid Systems: Computation and Control (2007). He is the recipient of several best paper awards at various conferences, and of an Advanced Grant from the European Research Council. Antonio Bicchi has been an IEEE Fellow since 2005.

Miroslav Krstic holds the Daniel L. Alspach chair and is the founding director of the Cymer Center for Control Systems and Dynamics at University of California, San Diego. He is a recipient of the PECASE, NSF Career, and ONR Young Investigator Awards, as well as the Axelby and Schuck Paper Prizes. Professor Krstic was the first recipient of the UCSD Research Award in the area of engineering and has held the Russell Severance Springer Distinguished Visiting Professorship at UC Berkeley and the HaroldW. Sorenson Distinguished Professorship at UCSD. He is a Fellow of IEEE and IFAC. Professor Krstic serves as Senior Editor for *Automatica and IEEE Transactions on Automatic Control and as Editor for the Springer series Communications and Control Engineering*. He has served as Vice President for Technical Activities of the IEEE Control Systems Society. Krstic has co-authored eight books on adaptive, nonlinear, and stochastic control, extremum seeking, control of PDE systems including turbulent flows and control of delay systems.